THE MODERN ENERGIES

Jenny Stanford Series on Renewable Energy

Series Editor

Wolfgang Palz

Titles in the Series

Vol. 1
Power for the World: The Emergence of Electricity from the Sun
Wolfgang Palz, ed.
2010
978-981-4303-37-8 (Hardcover)
978-981-4303-38-5 (eBook)

Vol. 2
Wind Power for the World: The Rise of Modern Wind Energy
Preben Maegaard, Anna Krenz, and Wolfgang Palz, eds.
2013
978-981-4364-93-5 (Hardcover)
978-981-4364-94-2 (eBook)

Vol. 3
Wind Power for the World: International Reviews and Developments
Preben Maegaard, Anna Krenz, and Wolfgang Palz, eds.
2013
978-981-4411-89-9 (Hardcover)
978-981-4411-90-5 (eBook)

Vol. 4
Solar Power for the World: What You Wanted to Know about Photovoltaics
Wolfgang Palz, ed.
2013
978-981-4411-87-5 (Hardcover)
978-981-4411-88-2 (eBook)

Vol. 5
Sun above the Horizon: Meteoric Rise of the Solar Industry
Peter F. Varadi
2014
978-981-4463-80-5 (Hardcover)
978-981-4613-29-3 (Paperback)
978-981-4463-81-2 (eBook)

Vol. 6
Biomass Power for the World: Transformations to Effective Use
Wim van Swaaij, Sascha Kersten, and Wolfgang Palz, eds.
2015
978-981-4613-88-0 (Hardcover)
978-981-4669-24-5 (Paperback)
978-981-4613-89-7 (eBook)

Vol. 7
The U.S. Government & Renewable Energy: A Winding Road
Allan R. Hoffman
2016
978-981-4745-84-0 (Paperback)
978-981-4745-85-7 (eBook)

Vol. 8
Sun towards High Noon: Solar Power Transforming Our Energy Future
Peter F. Varadi
2017
978-981-4774-17-8 (Paperback)
978-1-315-19657-2 (eBook)

Vol. 9
The Sun Is Rising in Africa and the Middle East: On the Road to a Solar Energy Future
Peter F. Varadi, Frank Wouters, and Allan R. Hoffman
2018
978-981-4774-89-5 (Paperback)
978-1-351-00732-0 (eBook)

Vol. 10
The Triumph of the Sun: The Energy of the New Century
Wolfgang Palz
2018
978-981-4800-06-8 (Hardcover),
978-0-429-48864-1 (eBook)

Vol. 11
The Triumph of the Sun in 2000–2020: How Solar Energy Conquered the World
Wolfgang Palz
2020
978-981-4800-84-6 (Hardcover)
978-1-003-00086-0 (eBook)

Vol. 12
Solar Euphoria: The Rise of Photovoltaics to the Top
Wolfgang Palz
2023
978-981-5129-00-7 (Hardcover)
978-1-003-43866-3 (eBook)

Vol. 13
The Modern Energies
Wolfgang Palz, ed.
2026
978-981-5352-02-3 (Hardcover)
978-1-003-79730-2 (eBook)

Jenny Stanford Series on Renewable Energy — Volume 13

THE MODERN ENERGIES

edited by

Wolfgang Palz

JENNY STANFORD
PUBLISHING

Published by

Jenny Stanford Publishing Pte. Ltd.
101 Thomson Road
#06-01, United Square
Singapore 307591

Email: editorial@jennystanford.com
Web: www.jennystanford.com

British Library Cataloguing-in-Publication Data
A catalogue record for this book is available from the British Library.

The Modern Energies

Cover image: Courtesy of Foster + Partners.

ISBN 978-981-5352-02-3 (Hardcover)
ISBN 978-1-003-79730-2 (eBook)
DOI 10.1201/9781003797302

Contents

Dr Wolfgang Palz

Dr Palz (on the left) with his good friend Giuliano Grassi, a wine grower in Tuscany.

Wolfgang Palz is a German physicist and a EU Official (retired). He holds a doctorate in physics and photovoltaics from the University of Karlsruhe, Germany. For more than half a century, he has been a global leader of the development of solar energy and other renewables. As an official of the European Commission in Brussels, he managed PV development in Europe for 20 years. From 2000 to 2002, he was a member of an "Enquête Commission" of the German Parliament on the German energy perspective on the horizon 2050. Later, he was a consultant for the EU Commission on PV program for the underprivileged in Latin America. Recognized as one of the fathers of PV and wind power success in today's global energy markets, he has received several recognitions, including the International Solar Energy Society (ISES) Global Leadership Award in Advancing Solar

Energy Policy (2011), the 2025 European Solar Prize, and a call to join the Chinese World Wind Energy Academy established in late 2025. He is a bearer of an Order of Merit of the German Republic.

Some Milestones in the Author's Professional Life

- 1965, Univ. Karlsruhe, Germany: Diploma in Physics. PhD in PV
- 1966–1970, University Nancy, France: Maître de Conférences
- 1970–1976, Paris, CNES (French National Space Agency)
    - 1973, Paris, UNESCO, Congress organiser "The Sun in the Service of Mankind"
    - 1976, Paris, UNESCO, Book publication "Solar Electricity"
- 1977–2001, Brussels, Official, EU Commission
    - 1977–1998, Division Head, Renewable Energies R&D for Europe
    - 1977, Luxemburg, Launch of 1st EU PVSEC
    - 1993, Budapest, ISES World Congress, "Power for the World, a Global PV Action Plan"
    - 1998, EuropeAid, Coordinator EU renewable energy aid for Africa
- 1999–2002, Berlin, Bundestag German Parliament, Member "Enquête Commission" on Germany's energy by 2050
- 2003, Germany, Order of Merit
- 2004, Brussels, EU, expert contract with EuropeAid, coordinator of PV projects for the poor in Latin America
- 2006, Beijing, China, Co-organiser "1st Great Wall World Renewable Energy Conference"
- 2007–2009, Paris, ADEME (French Energy Agency) Member Scientific Board
- Swiss Solar Prize 2010

- 2010–2016, Paris, French Government Programme Investments for the Future, R&D and Innovation PIA-1,
- Member Commission Nationale des Aides CNA
- International Solar Energy Society ISES, Global Leadership Award in advancing Solar Policy, 2011
- 2018, Singapore, with Jenny Stanford Publishing: *The Triumph of the Sun*
- 2023, Paris, 50th Anniversary Congress, "The Sun in the Service of Mankind"
- 2025, European Solar Prize

Some of the 41 co-authors of the book Power for the World, by Wolfgang Palz, published with Jenny Stanford Publishing, Singapore, in 2010: all my friends at a fiesta in Valencia, Spain. Since 2010, Jenny Stanford in Singapore publishes his series of Solar books.

Introduction

An energy revolution is going on today that will benefit all of us, as it makes our basic needs in electricity more affordable and the world a cleaner place. PV and wind power are witnessing worldwide growth, be it gigawatt-level central power stations or building-integrated plants installed by millions of individuals. Wind energy has rapidly become a leader of new developments, and PV has achieved dominance in the global energy markets—financial benefits being the main drivers. The current decade has seen wind power and PV eventually reaching the top of the energy markets, with nuclear power's decline accelerating. The intermittency of PV and wind power is being rapidly overcome through massive integration of battery energy storage systems in various sizes.

This book is not about wishful thinking and speculations, but hard facts and figures. It has been written by international experts who have made this energy revolution possible. Scientists from China, today's powerhouse in new energies, have made new information available exclusively in this book.

The book provides an overview of the global development of renewable energy from its beginnings in the mid-1950s all the way to its leadership today. The objective views of experts presented with the global well-being in mind are of general interest, highlight the benefits of renewable energy, and give evidence of the lies and biased and distorted views of the defenders of the old conventional energy, which had its day. The book explains how, thanks to renewable energy, the low-income populations in the developing nations can at last get access to affordable energy. It demonstrates how the world has now understood the benefits provided by our friendly and inexhaustible sun power. We hope the book will be an interesting read for both experts and general readers.

We gratefully acknowledge the invaluable contributions by the authors, which made this book possible. The assistance of Arvind Kanswal from Jenny Stanford Publishing in the finalization of the book is appreciated.

Dr Wolfgang Palz
Editor

Chapter 1

Photovoltaic Integration in Architecture

Stefan Behling

Senior Executive Partner, Foster + Partners, London, UK

sbehling@fosterandpartners.com

In 2027, Foster + Partners will complete its sixth decade. Reflecting on an era that saw a continued drive for innovation alongside consistent design rigour, we are careful to balance the technical with the sensibilities of the users who, alongside performance, are central to our work.

This experience has also given us a unique perspective on the unfolding climate crisis. We know that relying on a technology-focused, consumption-reduction approach is not enough. It must be combined with an architecture rooted in the needs of people and planet—material and spiritual, measurable and intangible—and that promotes healthy, connected and socially fulfilling lifestyles.

Today, Foster + Partners hones and applies its thinking to a variety of projects that greatly vary in scale and energy requirements. Being an international practice, these projects span geographies, climates, and countries, each of which has varying site conditions and infrastructural circumstances and, therefore,

The Modern Energies
Edited by Wolfgang Palz

ISBN 978-981-5352-02-3 (Hardcover), 978-1-003-79730-2 (eBook)
www.jennystanford.com
DOI: 10.1201/9781003797302-1

a diversity of suitable forms of generation that might be implemented. The particular focus of the following article is solar energy, though this is not to deny the importance of wind, hydroelectric, geothermal or, indeed, nuclear, in the pursuit of clean energy and a reliable grid.

These projects are milestones in responsible, technical and sustainable design that respond to and revitalise their places with the assistance of photovoltaic technology. The Reichstag (1999) signalled a national endorsement of renewable energy. At Masdar, in Abu Dhabi (2015), the new city quarter recreates in contemporary—and sustainable—form some of the essential character of historic Arab settlements. Bloomberg's headquarters (2017)—in what was London's Roman heart—adopts the stone solidity of its neighbouring buildings and reinstates part of a lost Roman road. Apple Park (2019) restores a vanishing Californian landscape of orchards and oak savannah and, alongside Bloomberg, achieves extensive natural ventilation in a deep-plan workplace—just one facet of environmental design we know to greatly benefit our health and well-being.

These projects have embraced on-site solar energy as part of their technological—and architectural—toolkit for reducing their environmental impact. To create a reliable grid structure that is prepared for renewable transition, the site of energy generation and storage, such as solar power within the fabric of architecture, must extend beyond the original, centralised grid.

Evidently, then, it must be early in the design process that architects and engineers work together and ask what opportunities there are to generate energy on or near a project site. This is best begun by using local climate data—that often spans 40 or 50 years—in combination with computer models that can account for climate change.

> *Is the site windy or sunny? Is there a potential for tidal or wave turbine energy? In all cases, the seasonality of a site must be considered. If a location has good wind potential in January, for example, but excellent solar potential in July, then ... 'generation shifting' might be suggested, which offers a mixed-modal approach to generating energy. In cases where*

energy potential from renewable sources is prohibitively low or excessively intermittent to be relied upon entirely, more stable clean generation technologies such as next-generation geothermal can be called upon to provide support. Even the newest small-scale nuclear generators have been considered, depending on project size and time scale. This is about designing in a way that is both site-specific and adaptable to changes in seasonality.

—Andreia Guerra Dibb and Jaymin Patel,
Foster + Partners Environmental Engineering

From out of this renewable energy tapestry came the opportunity to explore one of its threads—that of solar power—and interrogate how it has been woven into the past, present and future of sustainable architecture globally and at Foster + Partners. We can see how the uniquely flexible geometry and modularity of the photovoltaic panel have made it often the most specified renewable energy source at the architectural or urban scale.

1.1 A Solar History

The oil crisis of the 1970s was brilliant news for solar energy. Until then, photovoltaic panels (devices that convert sunlight into electricity) had been sidelined at the experimental, even extra-terrestrial fringes of energy generation. By 1975, however, increasing research and investment meant that the cost of these panels had fallen to under $20 per watt, down from $100 a few years earlier—a trajectory which promised to make solar energy cheaper and therefore more accessible to a wider market.

That same decade, photovoltaic panels were installed in remote, off-grid locations and certain bespoke, research-based architectural projects. In 1973, for example, researchers at the University of Delaware, United States, integrated them into a project named Solar One. This was a building that used heat and electricity converted from sunlight, and was one of the first of its kind.

Evidently, the drop in price and advance in technology had excited not only the energy sector but the architectural profession as well. The integration of photovoltaics into buildings and urban plans appeared, for the first time, a real, widespread possibility.

Figure 1.1 In 1973 and 1974, NASA pushed the boundaries of long-duration human space missions with Skylab, America's first space station. With three crews performing hundreds of science experiments and unprecedented observations of the Earth and the Sun, Skylab laid the foundations for the space science program on the International Space Station and for future missions to the Moon and Mars. *Source*: NASA.

Foster + Partners was alert to the developments and investments in solar technology taking place throughout the 1970s. In

1975, for example, the young practice completed a regional study for Gomera in the Canary Islands, which proposed several pioneering energy strategies. Energy self-sufficient houses were conceived to be supported by natural resources: wind power generates electricity for lighting and anaerobic generators digest household waste to create methane gas for cooking. The combination of almost constant sunshine and reliable winds on Gomera made the island a perfect testbed for an alternative energy-powered grid—with solar energy, in this case captured to heat water, being a key feature.

Figure 1.2 A frame from a sequence of sketches of the Gomera Case Study by Birkin Haward. The combination of almost constant sunshine and reliable winds made the island a perfect testbed for alternative energy sources. © Birkin Haward / Foster + Partners.

Figure 1.3 Drawing of a home on Gomera and how it could be supported by natural resources; wind power generates electricity for lighting; sea-water deliveries allow drinking water to be distilled in solar stills; and anaerobic generators digest household waste to create methane gas for cooking. © Pedro Guedes / Foster + Partners.

It can take many years between investment in the research and development of a new technology and its widespread adoption to take hold. This was the case for photovoltaic technology, where barriers such as unit costs, installation costs, lack of planning incentives, and inefficient energy storage remained limiting factors for the further adoption of solar cells into the built environment.

In the United Kingdom, the first solar panels were installed in 1978 at the Garden House at the University of Sussex; and the first domestic solar panels were installed on the Solar Park social housing in Woking, Surrey, in 1986. But it wasn't until 1994 that panels were connected to the national grid, in the 'Autonomous House' by Brenda Vale. A year later, in 1995, the Oxford Ecohouse, built by Professor Sue Roaf, became the first home in the United Kingdom with a fully integrated solar roof that connected with the grid. Though a bold and hopeful articulation of the future of energy-efficient homes in Britain, the Ecohouse was nevertheless still limited in its accessibility. The roof cost £28,000 (approximately

£58,300 today)—an expensive option for a regular homeowner—and was largely paid for by a supporting solar cell company. Twenty years after their introduction to the market in the 1970s, interest in photovoltaics had gradually broadened. Efforts to subsidise this technology and introduce it in a civic capacity were undertaken most notably in Germany.

1.2 The Reichstag's Solar Vision: Photovoltaic Roofs, the European Commission, and Drafts for the Solar City

In 1992, Foster + Partners were commissioned to design the Reichstag in Berlin. Following a set of high-profile international competitions, the practice was commissioned for a design that would later be completed in 1999 (after the former building had been *Wrapped* by Christo and Jeanne-Claude in 1995).

The Reichstag was not only a monumental restoration and renovation task but also a civic project that needed to represent a reunified Germany in the wake of a painful twentieth century. More so, as the new house for the Bundestag (German parliament), the building needed to look ahead to the twenty-first century, embracing its future technologies and philosophy of democratic leadership. Its transformation is rooted in four major issues: a belief in the significance of the Bundestag as one of the world's great civic forums; a determination to make the process of government more accessible; an understanding of history as a force which shapes buildings as well as the life of nations; and a passionate commitment to the low-energy, environment-friendly agenda which is fundamental to the architecture of the future.

The design team—Foster + Partners, Kaiser Bautechnik and Kühn Bauer und Partner, in conjunction with the German Federal Government and the European Commission—developed the brief for an energy-efficient building. A dramatically different architectural project than the smaller-scale, speculative research proposals of the Ecohouse, Solar One, or the Garden House, the Reichstag offered its designers an opportunity to bring photovoltaic technology to an international stage. These ideas converged in a structure that combines the manifestation of a German democratic government (which is world-leading in its legislation concerning environmental protection) and the potential for a sustainable, environmentally responsible public building.

Figure 1.4 The west façade of the Reichstag, New German Parliament, Berlin, Germany. © Nigel Young / Foster + Partners.

Photovoltaic cells, contained within 100 solar panel modules, were installed on the roof. The energy that these cells generate—approximately at peak output—is then harnessed to drive the exhaust air ventilation system in the public chamber (where the Bundestag sits).

The Reichstag's roof and glass dome is open to the public as an enactment of an egalitarian and democratic political process. One consequence of introducing the public to the roof is that these solar panels were made as visible as the political processes of the Bundestag; the engineering of the building and its endorsement of a solar architecture are, therefore, celebrated and made explicit, becoming an integrated architectural feature as well as a political endorsement for renewable energy.

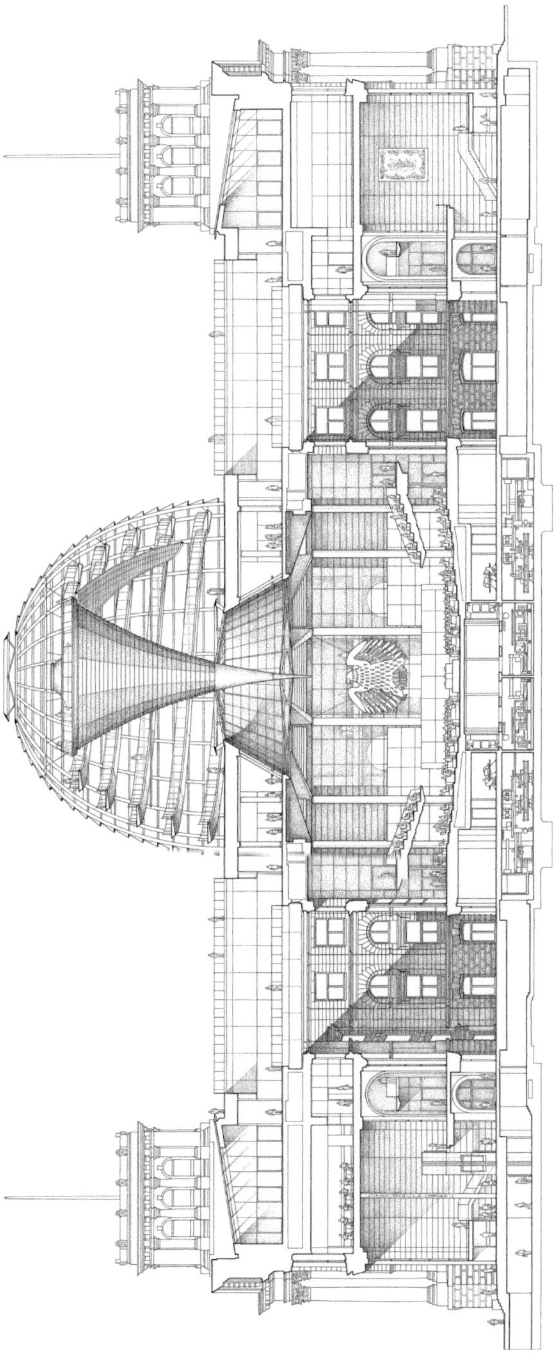

Figure 1.5 North-south cross-section through the chamber at the heart of the Reichstag. © BPR for Foster + Partners.

Figure 1.6 Solar panels on the roof of the Reichstag, seen from above and below. © Nigel Young / Foster + Partners.

At that time, programmes for a more widespread, lower-cost form of solar architecture were also being explored in Germany and the European Union. Political figures such as Hermann Scheer—following a subsidy programme set up by Walter Sandtner, Division Head for Renewable Energies in 1989—introduced the '100,000 Roofs Programme' (*100.000-Dächer-Programm*), which aimed to build 100,000 solar residential roofs in Germany, as well as the 'Feed In Tariff' in 1999, which paid residents for the kWhs that their buildings delivered to the grid.

During the design and construction of the Reichstag, and alongside the emerging political support for the integration of photovoltaics into buildings, Foster + Partners also leveraged the support of the European Commission, in collaboration with Wolfgang Palz, to extend the research into the potential of solar energy at an urban scale. In 1995, the study for a 'Solar Quarter' on an extensive inner-city site of Regensburg, including the island of Unterer Wöhrd in the Danube, emerged from the newly founded Renewable Energies in Architecture and Design (READ) Group chaired by Norman Foster and the European Commission.

> *'It was felt that the chosen city should have at least 100,000 inhabitants and should have a rich architectural, urban, social and cultural background. It was also felt that the location in this particular study should not be a greenfield site but rather be a brown/grey site within the existing urban fabric, helping to set a sound basis for a sustainable development.'*
>
> *'In particular, simulation techniques to study the effects of orientation have proved a most useful tool in quickly understanding the impact of fundamental design decisions. The ability to explore the potential of combined heat and power systems for the clearly defined area of the island have led to an understanding of how buildings can work together. The interest and involvement of specialists in the field of monitoring and control systems has allowed us to explore the potential these rapidly developing technologies can have on the increased efficiency of our cities. Finally, the ability to use new technologies, highlighted by our contact with specialists in the field of bio-technology, in our cities illustrates how we can turn waste into gain.'*
>
> —From the Solar Quarter Application

Figure 1.7 Drawings and models made as part of the European Commission Grant for Linz Solar City. © Foster + Partners.

Alongside Regensburg, a plan for a 'Solar City' in Linz, Austria, was also considered. This research focused on how solar technology integration could be achieved within the constraints of a standard social housing budget. The designs also address issues of standardisation in the building module while also allowing for maximum flexibility, to enable modification to suit individual circumstances.

> *'Solar architecture has largely concentrated on individual building components such as solar and photovoltaic panels, sun-shading devices and thermal storage elements. Some advances have been made in integrating these elements into the design of individual buildings and thus reducing their energy requirements.'*

The 1995 Solar Quarter and Solar City studies captured much of what the practice has since integrated into its design tools. In 2024, for example, Foster + Partners' Applied Research + Development group released Cyclops, a freely available design plugin that allows designers to turbocharge ray-tracing simulations that are fundamental for environmental analyses. Cyclops can run environmental analyses that used to take hours, or even days, in a matter of seconds, allowing designers to see the effects of building form and massing on ray tracing (sunlight) in real time. For an architect wanting to integrate photovoltaics into their design, or measure how sunlight works at an urban scale, such tools are essential in optimising sunlight exposure and ascertaining where to place panels and buildings for maximum efficiency and, therefore, improved sustainability.

1.3 Integrating Photovoltaics in Bishops Square

At the turn of the century, photovoltaics continued to be incorporated into projects by Foster + Partners—despite a slump in the market caused by investment bubbles in Germany, Italy, and Spain 'bursting.' Appointed to the practice in 2001 and completed in 2005, the Bishops Square development in London was, at the time, the largest photovoltaic installation in Europe. The project saved 23 tonnes of CO_2 by generating 54,000 kWh annually through photovoltaic panels fitted on the modest but important roof garden.

The Bishops Square projects proves that commercial buildings *can* integrate photovoltaic technology relatively straightforwardly onto their roof space, leading to a reduction in carbon-intensive energy consumption that is beneficial not only for sustainability but for operational costs. It also demonstrates how commercial buildings can become leading examples of how photovoltaic technology can be integrated in city life; solar architecture need not be confined to the experimental homes of the 1990s or the global icons of government.

Figure 1.8 Solar panels and roof gardens on the roof of the Bishops Square development. © Nigel Young / Foster + Partners.

1.4 Realising the Masdar Institute: A Solar-Powered City

Photovoltaics were a key consideration in powering Masdar City, a masterplan project appointed to Foster + Partners by the Masdar-Abu Dhabi Future Energy Company in 2007. The 640-hectare (1,580-acre) project is a key component of the Masdar Initiative, established by the Government of Abu Dhabi to advance the development of renewable energy and clean-technology solutions for a life beyond oil. The broader city masterplan combines state-of-the-art technologies with the planning principles of traditional Arab settlements to create a desert community that aims to be carbon neutral and zero waste. Solar power was deemed a viable source of energy due to the city's location in the United Arab Emirates—a country that receives strong-to-intense annual solar radiation for 8 to 11 hours a day.

Figure 1.9 Visualisations of Masdar City, Abu Dhabi, UAE. © Foster + Partners.

Figure 1.10 Rooftop photovoltaics at the Masdar Institute, 2010. © Nigel Young / Foster + Partners.

A mixed-use, low-rise, high-density development, Masdar City includes the headquarters for the International Renewable Energy Agency and Masdar Institute. Strategically located for Abu Dhabi's transport infrastructure, Masdar is linked to neighbouring communities and the international airport by existing road and rail routes. With a maximum distance of 200 metres (656 feet) to the nearest rapid transport links and amenities, the city is designed to encourage walking, while its shaded streets and courtyards offer an attractive pedestrian environment, sheltered from climatic extremes. The land surrounding the city contains wind and photovoltaic farms, research fields and plantations, allowing the community to be entirely energy self-sufficient. Knowledge gained in design development has already aided the formation of Abu Dhabi's 'Estidama' rating system for sustainable building in the region.

Part of this vision had already been realised in the Masdar Institute, a key phase in the development, which primarily completed in 2010. The Institute embodies the principles and goals of Masdar to create a prototypical and sustainable city and is the first district of its kind to be powered entirely by solar energy. The design incorporates a variety of passive and active environmental strategies and will be used as a testbed for the sustainable technologies that will be explored in future Masdar City buildings. Photovoltaic cells were fitted on the roofs of the buildings, and 3,156 solar panels were installed in a solar farm nearby—providing up to 10 megawatts of power to the site.

Figure 1.11 In many areas, the PV panels overhang the roof to provide protection from the sun. Other vernacular elements were introduced to Masdar's design, including narrow streets and screens for shading. © Nigel Young / Foster + Partners.

Figure 1.12 The Masdar Institute solar farm, completed in 2009. © Oliver Jackson.

While long-lasting and strong solar radiation is excellent for photovoltaics, excessive radiation is not advantageous for people's well-being. In the design of Masdar City, Foster + Partners needed to balance the potential for solar energy with the community's need for shade and for urban cooling.

Calling upon the lessons of vernacular architecture from the UAE, which developed architectural forms that could both harness sunlight *and* minimise its effects, the design team created a low-energy urban cluster by prioritising local climate and concerns and passive design strategies. The Masdar Institute, for example, was built with narrow streets to provide shade for pedestrians, strategically oriented buildings to block hot winds, shaded façades using external screening devices, and thick-walled buildings to enhance thermal mass and store heat. Courtyards are designed to offer shade and allow controlled air movement, and well-shaded water features are integrated to benefit from evaporative cooling which passively reduces ambient temperatures.

1.5 The Largest Building-Integrated Photovoltaic Installation: Apple Park

Apple Park, which completed in 2017, is the headquarters of Apple in Cupertino, California. Its design places people at its heart, creating an ideal workplace for continued creativity, innovation and well-being. The campus is Californian in spirit: open and connected to nature and the product of a remarkable collaboration between Apple and Foster + Partners.

Figure 1.13 Aerial visualisation and photograph of Apple Park. © Foster + Partners.

The simple form of the Ring Building conceals immense expertise and innovation. It comprises a few core elements: communal 'pod' spaces for collaboration, private office spaces for concentrated work, and broad, glazed perimeter walkways—featuring the largest sheets of curved glass ever constructed—that allow uninterrupted connection to the landscape.

Figure 1.14 A walking trail circumnavigates the outer façade of the Ring Building at Apple Park. © Nigel Young / Foster + Partners.

One of the most advanced precast concrete structures in the world, over 4,000 slabs, which span up to 15 metres (48 feet), make up the Ring's floors. Known as 'void slabs', these multi-use elements form the structure and exposed ceiling, incorporate radiant heating and cooling and provide air return.

The landscape and building of Apple Park form a seamless whole: the Ring Building, Steve Jobs Theater, Fitness & Wellness Center, Visitors' Center and South Parking are all encompassed by flowing parkland that enhances the buildings as places to socialise, exercise and work. Flexible and forward-looking, the architecture sits low amid tall trees and brings the invigorating views and fresh air from the park through its glass façades.

Figure 1.15 Solar panels under construction on the roof of Apple Park and the nearby South Parking infrastructure. © Nigel Young / Foster + Partners.

The site draws its energy from the sun with 45,170 solar panels, covering 60,400 square metres (650,000 square feet), fitted across the site—namely atop the Ring Building and the South Parking. As Apple reports, it is 'now the largest LEED Platinum-certified office building in North America. It is powered by 100 per cent renewable energy from multiple sources, including a 17-megawatt on-site rooftop solar installation and four megawatts of biogas fuel cells, and controlled by a microgrid with battery storage. It also gives clean energy back to the public grid during periods of low occupancy.'

Figure 1.16 Solar panels blend into the car parking infrastructure at Apple Park. © Nigel Young / Foster + Partners.

On-site batteries were vital, as the storage of energy is equally important as its generation. Though renewable energies, including solar, have become far more efficient and cost-effective technologies over the past decade, they cannot guarantee a constant supply (solar, clearly, only works during daylight hours). Designing in ways that reduce pressure on a national and a local grid as it attempts to transition to renewable power must, therefore, factor in storage systems that can reabsorb and store energy for hours, days, perhaps even weeks. Like the data centres that house our ephemeral 'cloud' data, energy storage is built, requiring both materials and spatial planning. As energy demands increase, greater storage capacity must also be increasingly integrated into design.

In addition to its solar aspirations, the 71-hectare (175-acre) Apple Park site—which was previously dominated by impervious surfaces—was transformed into a green space, at an increase from 20 to 80 percent, with over 6 kilometres (4 miles) of walking and jogging trails. There are now over 9,000 trees on site, including indigenous oaks and orchards, as well as meadows, sports fields, terraces and a secluded pond. Counter to the view that solar panels—and their accompanying storage needs—encroach on green space, Apple Park proves that with careful design, ecological and energetic sustainable planning *can* co-exist.

1.6 Kuwait International Airport: Responding to Innovation in Photovoltaic Technology

Advancements in solar technology were also absorbed into the design story of Kuwait International Airport (KIA), a project commissioned the same year as Apple Park. The airport is planned to significantly increase capacity and establish a new regional air hub in the Gulf—these strategic aims will be matched by a state-of-the-art terminal building that will provide the highest levels of comfort for passengers and will set a new environmental benchmark for airport buildings. Its design is rooted in a sense of place, responsive to the climate of one of the hottest inhabited environments on earth, and inspired by local forms and materials.

The terminal has a trefoil plan, comprising three symmetrical wings of departure gates. Each façade spans 1.2 kilometres (0.75 miles) and all extend from a dramatic 25-metre (82-foot)-high central space. The terminal balances the enclosure of this vast area with a design that is highly legible at a human scale—for simplicity and ease of use, there are few level changes. To further aid orientation, the building is planned under a single roof canopy, punctuated by glazed openings that filter daylight, while deflecting direct solar radiation. The canopy extends to shade a generous entrance plaza and is supported by tapering concrete columns—their fluid, organic forms draw inspiration from the contrast between the solidity of the stone and the shape and movement of Kuwait's traditional dhow sailing boats.

Figure 1.17 Visualisation of Kuwait Airport, scheduled to be completed in 2027. © Foster + Partners.

Drawing on the region's culture of hospitality and welcoming guests to Kuwait, the design establishes an elegant and memorable arrival sequence for passengers, which includes a baggage reclaim area surrounded by cooling cascades of water. The design also features a grand new landside access sequence from the south. Close to the building, the landscaping is a lush oasis, with strands of drier planting and species native to the desert climate extending further away from the terminal.

Due to the impressive scope and scale of the project, Foster + Partners developed a design for a roof with 132,000 photovoltaic panels, and 3,000 thermal panels in 2010. This would cover an area of 160,000 square metres (1.7 million square feet)—almost 20 football fields. In the 15 years since this submission, the materials, methods, and technology of the panelling proposed have greatly advanced. One core principle of Foster + Partners' design approach is to anticipate and adapt to these new technical developments. The original plans submitted outlined photovoltaics that would generate a minimum of 20 gigawatt hours of solar energy. Today's photovoltaic technology means that the same amount of energy can be generated by half the area, a statement that *could* be interpreted to mean that only half the proposed panels were needed to meet initial requirements. After careful negotiation, the original and full number of panels was agreed upon, meaning that an increased supply of renewable energy—far more than the 20 gigawatt hours proposed—will be achieved. This, combined with other strategies (including the thermal properties of the concrete structure), is part of KIA's aim to attain LEED GOLD environmental accreditation—the first passenger terminal in the world to do so.

1.7 The Future of Solar Architecture

To step into the future, we must acknowledge the efforts of those in the past. The tireless advocacy of figures such as Wolfgang Palz has meant that solar energy is no longer the 'fringe' topic that it was in the 1960s, but part of an actionable solution to our energy crisis today. What's more, thanks to Palz and his complementaries' campaigning across international forums, the unique ability of photovoltaics to integrate into architecture has been closely studied and trialled.

These developments have been reflected on a practical level in the change in price of photovoltaic technology.

In the past 20 years, the price of photovoltaics has dramatically fallen. The international spot price for photovoltaic modules was 22 cents per watt in 2022. It was 12 cents in 2023, and then 9 cents in 2024—compared with $20 per watt in 1975. This is a 99.55 percent reduction in price, or a 0.45 percent comparable cost to fifty years ago, and the technology is more sophisticated than ever. Today, the price for a complete photovoltaic rooftop system in Europe can be as low as €1,500 ($1,748) per kilowatt. Similarly, battery storage in combination with photovoltaics is now overtaking the battery market for electric vehicles.

This drop in price aligns with a rapid increase in installation. As of 2024, over 2 terawatts of photovoltaic technology have been installed worldwide—matching the world's coal-fired capacity. This second terawatt took a little over a year to install after the first terawatt, which was achieved in 2022. It is projected that by 2030, 8 terawatts in photovoltaic technology might be installed worldwide. This would mean that solar power could generate a significant portion of our global power supply.

The construction industry is responsible for 37 percent of global greenhouse-gas emissions. To move forward in line with climate goals, buildings and infrastructure need to be more efficient and sustainable in both their construction and operation. They must also provide people with the healthiest and most pleasant environment possible. Undeniably, the energy that is needed to achieve this should come from renewable sources.

Solar energy's dramatic fall in cost and rise in efficiency make it an essential tool, but never the only one. As Foster + Partners' work shows, solar must be understood within a wider tapestry of renewable solutions, always designed in response to people, place and climate.

The future will not be defined by a single technology but by the intelligent interplay of many: photovoltaics embedded in façades and rooftops, wind and geothermal systems harnessed where the site allows, and new forms of nuclear supporting stability and resilience. Not to mention the need to continually evolve our understanding of embodied carbon in construction, and life-cycle carbon in the operation of our architecture and infrastructure. For architects, the task is to shape these forces into environments that answer not just technical needs but human ones. In this sense, solar is a reminder that sustainability must be both measurable and intangible—rooted in material performance yet also in the qualities of life, connection and well-being that buildings can nurture.

Chapter 2

The Modern Energies: A Global Overview

Wolfgang Palz

Official (ret), EU Commission, Paris, France

w@palz.be

This chapter is dedicated to

My wonderful father, Karl J. Palz (1892–1982),
German Professor of Mathematics, Astronomy, and
Natural Sciences

And

My sons,
Christophe Palz, Strasbourg, France,
and
Alexander Carlo Palz, Brussels, EU Capital

The Modern Energies
Edited by Wolfgang Palz

ISBN 978-981-5352-02-3 (Hardcover), 978-1-003-79730-2 (eBook)
www.jennystanford.com

DOI: 10.1201/9781003797302-2

Figure 2.1 The author with Mona Neubaur, Deputy Premier and Minister for Economic Affairs, Industry, Climate Action and Energy of the State of North Rhine-Westphalia, when he received the European Solar Prize 2025 in Düsseldorf, Germany, that year.

Figure 2.2 Building a great solar wall against a world in distress, climate degradation, wars, and poverty.

2.1 Introduction and Summary

2.1.1 The Breathtaking Rise of Solar PV, Wind Power and All Renewable Energies

In 2025, PV is number 1 of all globally installed electric power providers—matched only by a small margin by coal, the traditional leader. For the past few years, the development of PV has accelerated exponentially.

Almost all countries around the globe are participating in PV's surprising rush. Millions of people have invested in it and run their own solar power plants.

Today's PV success was unthinkable without China's push! China is the leader of global PV deployment. It leads industrial development, and its domestic market is the world's biggest. Most PV modules employed in the world today come from China.

Today, China is the world leader of all renewable energies, including PV, wind power, hydro power, bioenergy. It was not always like that. This book will explain how this happened.

Renewables stand for a new business, a clean business, with hundreds of billions of dollars invested every year in new plants and operations.

Solar energy has created millions of jobs over the past few years.

A Solar Revolution: Overall, today the renewable power capacity already contributes 4.8 TW (4.800 GW), some 50% of the total global capacity, which still includes the fossil and nuclear power.

As already said, PV comes on top with well over 2.8 TW installed in late 2025, followed by wind power with over 1.3 TW. Hydro power contributes the same share as wind power, followed in size by bio-power, geothermal power, and marine power.

In terms of global electricity production from this power, in 2024 renewable energies made a contribution of over 35% of the total—in Europe about 50%. With over 14% of the contribution, hydroelectricity comes first, followed by wind electricity with over 8%, and PV with some 8%, too.

Important note: The renewables industry is doing well! In the first half of 2025, the Dow Jones "S&P Global Clean Energy Transition Index" grew by 25%, with PV coming on top. The period coincides with the first 6 months of Trump in office. It is just a coincidence, certainly not Trump's merit; the Index is global and not limited to the United States. A major contributor is China.

The other revolution comes with the development of battery energy storage system (BESS), associated with the renewables on a large scale. Thus, what used to be the intermittent power of PV and wind is converted to a stable electricity supply.

Thanks to its exceptional growth, PV will have passed all others by 2030 and will double its contribution to electricity generation to over 17%.

According to Bloomberg's BNEF the world has invested $386 billion in renewable energies in the first half of 2025, 10% more than in the first half of 2024. China made up 44% of the total.

The EU invested $30 billion in the first half of 2025, 63% more than the previous year—with a particular increase in off-shore wind power in the North Sea.

Trump's USA invested this time 36% less in the renewables, only $20.5 billion.

The global off-shore wind power got $39 billion in the first half of 2025, more than the total of $31 billion it secured in the whole 2024.

Solar PV increased as well but showed a mixed picture: Small-scale PV increased very much; in China it even doubled. Utility-scale PV decreased globally by 19%.

2.1.2 Further Growth Ahead on the Road to Come: Up from Today's Share of the Renewables in the World's Primary Energy Use

In early 2025, renewable energies have contributed over 50% of the global power capacity and 35% of electricity generation—in 2000, it had been only 19%, as hydro as it were. Since 2000, additional PV and wind power projects have been deployed, which now provide 16% of the electricity mix in 2025.

In terms of renewable fuels applied in industry, buildings, and transport, their share is not yet more than 5% of the total demand. Most of it is used in the form of modern biomass. Clean hydrogen as a fuel is still in its infancy.

Soon, by 2030, the share of the renewables in all sectors is expected to grow to 20%, which would be an overproportional increase over the total increase in energy consumption expected by that time: Energy consumption will further increase in the future, too.

The global electricity demand is even increasing faster than the total energy consumption as the world is electrifying.

Overall, oil today is still the leader in primary energy use with 32%, followed by coal at 26%, and natural gas at 23%.

2.1.3 Multiple Benefits from Solar Energy's Rise: A Key to Preserving Global Environment and Climate, Trimming Cost of Energy Supply, Reducing International Tensions

According to the International Monetary Fund (IMF), subsidies toward the use of fossil energies amounted to some $7000 billion in 2024. They did not decrease; rather they grew by $2000 billion in the last two years. Fossil fuels fire the vicious cycle of climate change: Their consumption raises the emission of CO_2 and other greenhouse gases (GHG); higher concentration of GHG raises the average global temperatures and a few other things such as air humidity, rainfall, floods, droughts, loss of biodiversity, and others, and the secondary consequences thereof.

The average temperature increase is more pronounced at higher Earth latitudes. The average scale of humidity increases with higher water temperatures in the oceans. The effects can change dramatically from site to site and time-wise. New York and the US East Coast, which used to be plagued by severe drought, received torrential rains later in the year 2025. With the torrential rains come floods.

The countries around the globe that were most hit by these phenomena recently were Pakistan, Taiwan, the Philippines, Nepal, Japan, Burma, Central Africa, the Sahel countries, Australia, Spain, the latter with over 200 deaths. China was much affected, too; the Three Gorges Dam was on the highest alert for flooding. Often the electricity supply is interrupted after inundation.

Agriculture is considerably affected by climate change and has to adapt, a major economic change that is.

Western Europe was wet in 2024 as never before. The Met offices in Germany, France, and Belgium were wondering that the rainfall in the year was the highest since they kept records—and solar radiation got lower accordingly. In 2025, it was the other way around: Central Europe was dry in the springtime and PV in the high production mood; on the other hand, Spain had weeks of rain, the Mid West of the United States received some torrential rains, too. It was as if the weather Gods wanted to make a new demonstration to the non-believers of the climate change effects.

And all these disasters add up to the economic and financial cost, difficult to get hold of in view of the diversity of the problem. The German Institute PIK on climate matters in Potsdam gave it a try. It evaluated the yearly damage through climate change globally to $38,000 billion. The authors concluded that the damages were 6 times higher than the costs needed to limit temperature increase to 2°C.

As a result of all this, in 2023 for the first time a record-high of 57 Gt of Green House Gases GHG (CO_2 + other climate killing gases) were globally emitted.

On March 21, 2025, the UNESCO warned that as a result of worldwide glacier melting, the supply of 2 billion people with water and food is endangered. Since 2000, 9000 gigaton of glacier ice have melted. Just in 2024, 450 gigaton got lost.

The World Meteorological Organisation WMO, in another warning, noticed that each of the last 10 years was the warmest in human history. In 2024, over 1.5°C above preindustrial times were reached. CO_2 concentration has been the highest in 800,000 years. NASA found that the global sea level increased in the past 30 years by 10 cm and the increase is currently accelerating.

Emissions are very different from country to country: Following UNEP, the UN Environment Programme, an authority in the field of environment and climate, in 2023 GHG emissions per head were highest in Russia with 19 ton, followed by the USA with 18 ton, China with 11, and the EU countries with some 7 ton. 55 countries in Africa emitted not more than 2.2 ton per head.

In a report issued in October 2024 in Nairobi, UNEP warned that at current policies the world will suffer a catastrophic temperature rise of up to 3.1°C—and climatic disasters on a gigantic scale, never seen before.

The UNEP report advocates the increased deployment of PV and wind power, having by 2030 a 27% total reduction potential of emissions, followed by the deployment of forestry for a 20% reduction potential. Other measures would include an increase in efficiencies, electrification, fuel switching in buildings, transport, and industry, it says.

In November 2024, the EU climate service Copernicus found out that 2024 was Earth's first year with an average temperature above 1.5°C compared to pre-industrial times, with 2.6°C increase being the long-term trend.

Following the IEA in Paris 37.4 Gt or 66% of the total GHG emissions, were energy related, in particular a result of the still increasing use of carbon-intensive fossil fuels. And to make things worse, today the fossil fuels are still subsidized to a large extent as mentioned above.

And the new US President is an unconditional defender of the use of fossil energies...

Just to mention them, food production (meat production in particular) and transport are two other important sources of GHG emissions.

There is general agreement that the renewable energies derived from the Sun are the obvious solution to save the world from a disaster to come. It is not nuclear power.

In 2024, $700 billion went on the investments for PV and wind power, $500 billion on PV and $200 billion on wind power, + $50 billion on battery storage, peanuts in view of the challenges ahead. And to be underlined should be the fact that most funding for PV and wind power deployment comes from private sources and not from State budgets, those that often move in the wrong direction, such as nuclear power and fossil fuel subsidies.

In a report from June 2025, the IEA in Paris, known for being a big defender of nuclear, could not but observe global Solar PV for coming in 2025 again as the number 1, the single biggest item of the $3300 billion energy investments in the year, with

$450 billion—and $65 billion for battery storage; a bit less so for PV than in 2024. And less than the $1100 billion to go for the fossil energies.

The good news is that 2024 was most probably the last year for which the GHG emissions were increasing on a global scale. Much electricity today is still produced from coal, the most important CO_2 emitter, but losing more and more competition in cost against PV, Wind electricity and consorts that don't produce the climate killers.

Climate change not only raises temperatures and destabilizes global weather but also increases pollution. The high pollution levels caused by ozone, particulates, and other effects in many parts of the world are further stimulated by climate change, be it in India, Pakistan, or Bangladesh. Even in the United States, the American Lung Association found in a report published in 2025 that 46% of the country's population is breathing unhealthy air, an increasing trend—Los Angeles being the worst with regard to ozone pollution and Bakersfield in California with regard to pollution by particulates.

The EU and the US decreased GHG emissions already in 2023, the EU by –8.3% and the US by a small –1.4%. And even better, the GHG emissions due to electricity and heat generation decreased by 24% in the EU in 2023 with respect to 2022. The fossil fuels still counted for 32.5% of electricity production there, but their share slowed by some 20% in 2023. The decrease in their use is triggered by increasing cost and the effect of the "EU emission trading," limiting CO_2 emission rights (see below). Instead, the share of the renewables in the EU's electricity production increased in 2023 by 12.4 % and attained 44.7% of electricity production; in 2024 it even did rise to over 47%. India and China, one of the world's two biggest emitters of GHG, still increased emission by 6%, resp. 5% in 2023. But since 2025, emissions have decreased in China as well.

There is an economic sense of renewable energy supply, too: unlike nuclear power, solar PV, wind power, and most other relevant renewable energies are cheaper than all others (a solar cell costs nowadays not more than a few cents per Watt). Millions of people who have already invested in solar energy have made the right choice. And a lot of politicians with their short-term view and lack of understanding go for the old conventional ones, as if nothing was learnt in the past 50 years.

The "Levelized Costs of Electricity" (LCOE) for PV and wind power were analyzed, for instance, by Wood Mackenzie, the competent analyst and consulting firm in Scotland, and published in October 2024. They were all better than those of the conventional fossil energies:

Ground-mounted PV in its simplest form on one axis were found 60$/MWh or 6 cents/kWh. It is an average value that can vary between some 3 to 10 cents/kWh depending on the technology and the provider, the local climate, the cost of capital, and others. In China, the cost of **distributed** PV came further down by 1/3 last year.

The typical LCOE for wind power in 2024 was 7.5 cents/kWh. Wind power being much site-specific it can vary between 2.3 cents/kWh and 14 cents/kWh. The offshore wind power costs came much higher: 23 cents for fixed turbines and 32 cents for floating plants.

In China, marine wind power is considered a premium technology. Just now, Dongfang Electric is producing a 26 MW turbine at Fushou, a world record. And so are its dimensions: a tower height of 185 meters and a turbine diameter of 310 meters.

Most solar energies are decentralized in use and are most suitable for the poor for getting access to a better life, and promote peace. No way to make bombs and missiles with PV!

Clean energies are with us to promote international co-operation against the disruptions of an international energy crisis. *Note*: Global tensions have never been as high in years as they are today.

2.1.4 The Carbon Markets: Other Instruments to Reduce Global GHG Emissions

The World Bank issued a report on the global carbon markets in June 2025, which notes that more and more countries, including Brazil, India, and Turkey, have introduced carbon taxes, and others like Paraguay or Vietnam are ready to do so. This helps to reduce their emissions and create more income to their national budgets. Such is supported green growth and the creation of jobs. It also helps mobilizing private capital to this end and orient those towards sustainable development.

Also, in June 2025, the French I 4 CE, the Institute for Climate Economics, published a detailed report on the subject. It found for the global income of the carbon markets in 2024 a total of $103 billion, an increase of over 50% from the corresponding market 5 years earlier in 2019. The report found that on May 1, 2025, 78 tarification instruments were in place worldwide, of which 35 markets of emission quota and 43 carbon taxes. So 65% of the world's GDP is part of the global carbon market. 28% of emissions are subject of a carbon tax. Are in particular concerned emissions from electricity and the industry. Agriculture and cattle breeding, big emitters, are so far excluded.

I 4 CE concluded that today's income generated in the global carbon markets is just a drop in the ocean compared to the needs of investments in sustainable development. Those are estimated between $6300 billion and $6700 billion per year.

2.1.5 The EU ETS, the Greenhouse Gas Emissions Trading System

The ETS requires polluters to pay for their GHG emissions. Launched in 2005, it is the first carbon market. It helps to bring down emissions and generates revenues to finance the green transition—subject of this book. It covers emissions from the electricity and heat generation, industrial manufacturing and aviation sectors, which account for roughly 40% of total GHG emissions in the EU. It started to cover emissions from marine transport since 2024. It operates in all EU countries plus Iceland, Liechtenstein and Norway.

By 2023, the EU ETS has helped bring down emissions from European power and industry plants by some 47% compared to 2005. Since 2013, the EU ETS has raised over €175 billion. The revenue flows primarily to national budgets. The Member states must use it to support investments in renewable energy, energy efficiency, and low-carbon technologies that help reduce emissions.

In 2025, the price is €67 per ton of GHG emissions.

2.1.6 ETS2

As part of a reform of 2023, ETS2 was created, separate from the ETS Directive. It covers the emissions from fuel combustion in buildings, road transport, and small industry.

ETS2 will become fully operational in 2027. BloombergNEF and the German ADAC have calculated that via ETS2 GHG emissions of 1 ton from the use of fossil energies could cost up to €149 by 2030, a doubling of its current price. This would mean 27% additionally for gasoline and up to 41% more for heating with oil and gas.

Accordingly, heating with oil or gas and driving a car running on gasoline would become a lot more expensive.

It will be the highest worldwide for households and small businesses.

2.1.7 The Chinese ETS

It was launched in July 2021 and included CO_2 emissions from the power generation sector. In March 2025, it was extended to involve steel, cement, and aluminum smelting. It will now also cover other GHG emissions, not just CO_2. After the expansion, the national ETS will cover 60% of the country's total CO_2 emissions instead of 40% before.

It will cover emissions from energy activities and industrial processes.

2.1.8 In the United States

The U.S. Environmental Protection Agency (EPA) is in charge of a Trading system. There are the emissions cap and tradable allowances.

Under the EPA's Power Sector Programs, first toxic pollutants such as SO_2 were targeted but not the CO_2 emissions. Those were the subject of new standards issued in April 2024 by the EPA. Between 2010 and 2023, emissions from coal-fired plants have been cut by over a half, and those from the gas-fired ones, the other leading fossil fuel-fired electricity provider, have increased. This trend corresponds to the actual electricity generated from such plants in the country.

Electricity supplied from natural gas doubled in volume, while that from coal decreased by almost 2/3 during the period.

2.2 Photovoltaics

The author chaired from the EU Commission in Brussels the European PV development from 1977 to 1998.

2.2.1 A Much Welcome Competitor: Solar PV Power, the Recent Star among All Energies

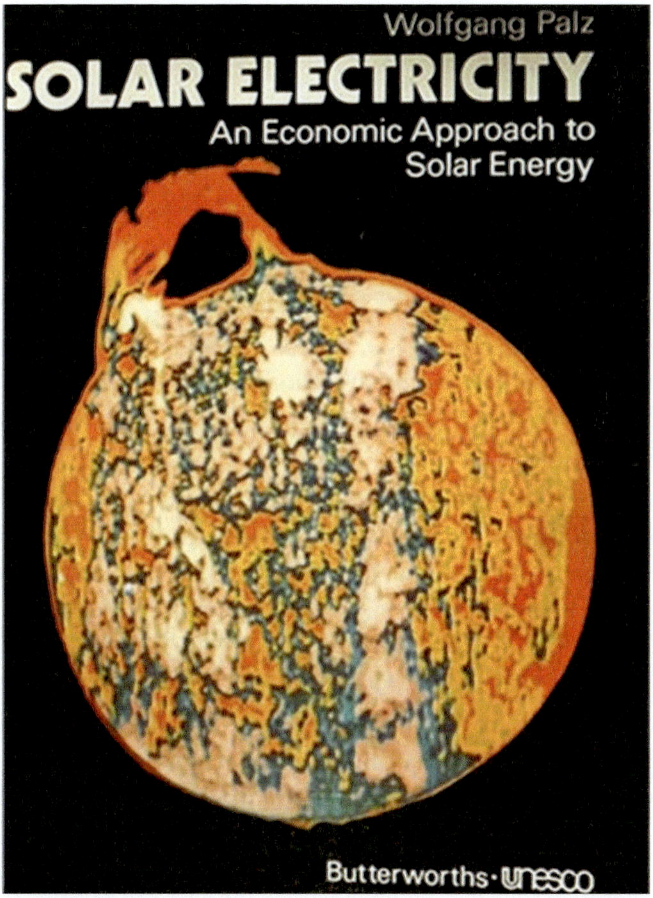

Figure 2.3 The author's book from 1976 on PV and other Renewables, published by UNESCO in Paris in seven languages.

Following the IEA in Paris was 2023, the 22nd year in a row that renewable capacity additions set a new record. As IRENA, the Renewable Energy Agency in Abu Dhabi put it, while Renewables' annual capacity increased by some 90% in 2024, non-Renewable capacity additions remained constant or even decreased, a tremendous contrast.

Solar PV alone accounted for ¾ of Renewable power additions in the world. One of the reasons is its low cost: PV's Levelized Cost of Electricity LCOE was in 2023 56% lower than that from the lowest-cost fossil fuels. Back in 2010 its cost was four times higher.

Confirmation comes from the US markets where 67% of all new power capacity additions to the grids in 2023 were from PV.

A global view of PV's recent developments was given in a paper of the prestigious IEEE Institute in the US (N. Haegel, S. Kurtz, IEEE Journal of Photovoltaics, Vol 13, Issue 6, Nov. 2023): "PV represented 61% of newly installed global electricity generating capacity for 2023. The amount of electricity generated by PV, Wind Power, Geothermal, and Bio-Power reached another record high and exceeded generation by global Hydro Power for the first time in history. Year to year growth in PV installations and PV generated electricity continued at remarkable levels (+35% and +24% respectively) while grid-scale battery storage grew even faster (+120%).

Newly installed global generating capacity in 2023 for combined Hydro, PV, and Wind was 87% of the total with PV at more than 50% for the second year in a row. The percentage of global electricity generation by PV was 5.5% for 2023. Furthermore the overall trend of a decrease of fossil generation capacity as a percentage of global new installations is confirmed.

Global PV capacity now slightly exceeds that of Hydro, although the higher Hydro-capacity factor means that Hydro Power remains the largest single source of Renewable electricity. If PV and Wind Power continue their recent growth trajectories, both are on track to reach and exceed levels of electricity generation from nuclear and Hydro in the next 3 to 5 years.

New energy storage is developing fast: Pumped Hydro Power has long been and remains the dominant energy storage technology supporting the electricity grid. Grid-scale battery energy storage systems BESS are only recently becoming a significant addition. Of the 55.7 GW of cumulative BESS reported, the largest installations are in China (49% at 27 GW), followed by the US (15.8 GW) and a combined capacity in Europe of 7 GW. Growth in battery storage in 2023 was in excess of 100%. For comparison, pumped Hydro capacities for the top three countries are China with 50 GW, Japan with 22 GW and the US with 17 GW."

2.2.1.1 Global PV deployment and its rise over the years

In early 2022, PV deployment achieved a major breakthrough: 1000 GW = 1 TW installed for the first time worldwide.

And end of 2024, just two and a half years later, the world had already *2 TW* installed.

Eventually, some 3 TW, close to 3000 GW will have been installed globally year-end 2025, most of it in China.

For the end of this decade by 2030 some 7 TW of PV are expected to be operational worldwide.

An early milestone was the installation of a first *1 GW* in the year 2000. From there it took 22 years to install the first TW and only some 2 years more to install the second TW. And now, later in 2025 already 2.5 TW are installed: PV's steep growth of deployment keeps accelerating!

Deployment of solar power has by now become of interest for all of us, all countries are becoming involved. In the rest of this chapter we're going to review PV progress just in a few of them, the major ones. Interestingly, even the "oil power" players of the Arabian countries that were in the early days strong opponents to solar that was seen by some as a new competitor in the oil business have come over as major solar investors.

More recently, Saudi Arabia, the UAE, Qatar, Egypt, Morocco, you name them, have started major new activities to promote solar power massively in their countries. Rightly so: they all got the solar resource, endless areas of cheap ground, and a lot of sand they are sitting on, that is exactly the material modern silicon solar cells are made of.

It all started in 1953–54: the semiconductor world of industrial silicon development. At Bell Labs in the United States, at the same time were invented that year transistor chips, on the one hand, and the industrial silicon solar cell, on the other.

The transistor, later followed by the IC, the integrated circuits, opened the door to a new world, the digital world in which we live since then. Silicon Valley in California became the breeding ground of cellular phones, Apple, Google, social media, more recently AI, and maybe more things to come.

Initially it was not clear why solar cells and PV power plants should not be suitable for electricity generation on a large scale, too.

The PV market for powering satellites has been identified as such early on: it makes sense to use the Sun's fuel available in orbit, thus avoiding high-cost fossil fuels for powering Earth's satellites.

For electricity generation on a large scale, that's another story. Since Faraday in London in 1830—and others that came later, like the German Werner von Siemens—the world knows that for generating electricity one employs a dynamo. All the known generators, classical or even wind power use a dynamo, but PV does not. That one has only to transform the DC current generated by the solar cells to AC on the grid lines.

And to those who claimed why it shouldn't work on a large scale as PV works so well for satellite generators, the reply was that for that purpose PV costs were very high.

The American pioneers in the early 1970s around Bill Cherry from NASA, the German US immigrants Martin Wolf, Karl Wolfgang Böer, then Paul Rappaport, the founder of NREL, the National Renewable Energy Laboratory, my friend John Francis Jordan and many others had a clear view of the PV solar world that was for all of them the future electricity of choice.

May I recall here Bill Cherry's wakeup call from 1971, as it is still fully valid today: "The large-scale utilisation of solar energy will be a legacy for generations to come, something for all citizens to be proud of and a major step towards cleaning our planet both from particulate and thermal standpoint. For these reasons, the large-scale utilization of solar energy should be initiated. It just may be the difference between survival and self destruction of man."

But somebody had to pay the bill to make solar energy competitive through mass production. And it was not America.

In Germany, a subsidy for PV installations on grid-connected buildings started in 1989 with the "1000 roofs PV programme." It had been initiated by my friend Walter Sandtner, the Division Head for Renewable Energies at the Ministry for Research and Technology, BMFT, in Bonn. It was the world's first, and others followed, namely in Japan.

Japan, following indeed Germany's first step, opened the mass market of solar PV since 1994, by promoting PV on grid-connected houses. Earlier on, 99% of PV's market was for isolated applications such as oil rigs. Stimulation in Japan came through government regulation for grid-connected PV, a subsidy, and net metering, another new tool for PV deployment.

In Germany was initiated in 1999 an extension of Sandtner's PV roof programme, Hermann Scheer's 100,000 roofs programme. It was followed, right at the turn of the century by the "EEG," when H. Scheer, together with H. J. Fell and other colleagues of the German Parliament was successfully introducing something new for PV deployment, the FIT, the "Feed In Tariff," now called EEG, that pays for the kWhs delivered to the grid, not the kWs installed. More details on it later.

Hermann Scheer's initiatives worked, and the PV markets started exploding! The early paymasters for the needed mass production became the Germans, the Italians, and Spain.

It ended as outlined in the previous book *Solar Euphoria* with the bankruptcy of all the flourishing German PV industry, not to say the whole system; the ever-increasing expenses reached astronomical dimensions and were stopped.

Progressively, China took over. Since 2015, China has acquired the leadership of global PV.

In summary, the world's PV markets went through many orders of magnitude in cost and volumes before becoming the energy wonder of today:

- **1953–54**: realization of the world's first industrial silicon solar cell in the United States, patented by Bell Labs and made widely known in the media.
- **2000**: the first 1 GW of PV in total deployed on Earth under Japan's leadership.
- **2009**: 10 GW deployed globally under the German leadership.
- **2012**: 100 GW deployed under the German leadership with Italy and Spain.
- **2016**: 300 GW globally achieved in total under the first year of the Chinese leadership.
- **2022**: 1000 GW, 1 TW, achieved globally under China's leadership.
- **2023**: the first year in which China installed more new PV at home than the rest of the world combined.
- **2024**: over 2 TW in total of solar PV capacity installed worldwide, the same as the world's coal-fired power capacity.
- **2030**: globally over 7 TW PV deployed being expected—over half of the Power capacity installed worldwide.

By **midyear 2025, 380 GW** of PV had been newly installed in the first 6 months worldwide, 64% more than in the first HY 2024, according to the consultancy Ember. The global total stood by July 2025 at 2.6 TW, and 3 TW are expected for the year-end 2025.

China installed in first HY 2025 256 GW more than twice what it had installed in that period the year before and twice the 125 GW of PV installed in the rest of the world combined.

Next to China, the following countries took the lead for PV in the first HY 2025:

- **India: 24 GW**
- **United States: 21 GW**
- **Germany: 7 GW—European leader**
- **Brazil: 7 GW**

The value of the global PV market in 2025 is evaluated at $200 billion.

The global module manufacturing capacity is to hit 1.8 TW/y.

Conclusion: What the early pioneers had anticipated has been achieved today. At least the first phase, until in the not-so-distant future, all our energy needs will be covered by solar energy, all of them.

The question was arising what could be the future for the renewable energies when the new US president comes to power: the day Trump was winning the election in November 2024, the losers of the day on Wall Street and the stock markets in Germany as well were indeed the renewables. Eventually the solar communities turn attention to Elon Musk, Trump's influential advisor who is committed to solar energy and storage hoping for support. It is said that Trump is a fan of oil but the fact is that Musk's Tesla is the champion of getting rid of oil-consuming cars.... **All wrong: Eventually late in April 2025, Trump was indeed swinging the hammer: up to 3000% (not 30 or 300%) taxes on PV module imports into the United States from Southeast Asia.**

And there are more trade tensions expected to come between the United States and China, the renewables' champion. Future will show...

Figure 2.4 The author's book on renewables, in Chinese version, presented in Beijing. (Right) Li Jufeng, Director of China RE Association and DG of National Center of Climate Change. (Left) Gao Jifan, the chairman of Trina and the Chinese PV Industry Association, the author and Qin Haiyan, the Secretary of the Chinese Wind Energy Association.

2.2.1.1.1 China's PV success, its industrial PV capacities

Since 2016, the world's leading PV module industry, with a few minor exceptions, is Chinese: and it is by and large profitable.

By mid-2024, details of the shipment ranking of the leading module manufacturers were released for the first half year H1 2024. Given below are some details for the biggest five of them:

- Jinko solar, shipped 44 GW in H1 (revenue $65 billion); profit $1.7 billion
- JA solar, shipped 38 GW (revenue $52 billion); loss $1.2 billion
- Trina solar, shipped 34 GW (revenue $60 billion); profit $0.7 billion
- Longi, shipped 31 GW (revenue $54 billion); loss $7.3 billion; despite its net loss, Longi kept its AAA bankability rate in 2024
- Tongwei TW solar shipped 19 GW (revenue $61 billion); loss $4.4 billion

The 10 biggest global manufacturers shipped 226 GW in H1, over 2/3 of it from the aforementioned five leaders. For a few years, profits decreased or even turned for some of them into a loss. The reason is the significant drop in international prices of PV modules due to excess production capacity of all parts of the supply chain.

In the full year 2024, globally some 600 GW of PV modules were produced and sold. The market leaders were Chinese like all the preceding years, but with a few newcomers:

- Jinko: 95 GW
- JA solar: 78 GW
- Trina solar: 78 GW
- Longi: 78 GW
- Tongwei: 45 GW
- Astroenergy: 40 GW
- Canadian solar: 30 GW
- GCL SI: 25 GW
- DAS solar: 25 GW
- Risen Energy: 22 GW

Other relevant manufacturers from outside China were Adani Solar in India, ELiTE and Maxeon in Singapore and VSUN in Japan.

From 22 cents/W in 2022, the international spot price for PV modules has decreased to 12 cents in 2023 and some 9 cents a year later. That's not much more than the cost of the glass for the encapsulation. It's pure magic!

In contrast, the technology of solar cells and PV modules was more sophisticated than ever. In 2024, n-type crystalline silicon TOPCon had a 70% share of the market. The share of the rest was mainly for PERC cells.

Longi was proud to announce a new record of 25.4% efficiency for its back contact HPBC 2.0 technology in the fall of 2024. Already in 2007, SunPower in the United States had reached a record efficiency of 20.3% with a BC cell, a record since then beaten many times. The efficiency of Longi's record-breaking solar cell was certified by a German FhG (Fraunhofer Gesellschaft) Institute. And later that year, Longi came with a record of 27% for an HBC cell.

By another FhG Institute in Germany was certified also at the year-end 2024 a world record of 25.9% efficiency for a Trina PV module with the dimensions 210 × 182 mm^2 employing n-type bifacial TOPCon cells.

Competition in announcements...

2.2.1.1.1.1 *China's PV home market*

China is the world's solar power house, also with respect to PV deployment at home. It did not happen by accident, but rather by government policies to make PV in China great. A 100 % success.

Early stimulation and support measures of China's market and the important changes after 2010 can be summarized as follows:

- China's PV industry had developed since the turn of the century for a purely export market.
- PV deployment in the home market started in earnest in 2011; since then, installation rates in China increased all the time. The FIT had been adopted for support in all the years, but it was abandoned in 2019. It was felt then that measures had to be taken to advance PV in a better way in the country.

In the years 2016 and 2017, there was a high curtailment and problems of grid connection. Expenses for FIT payments were getting too high. So, accommodating measures for increasing shares of PV had to be taken.

Figure 2.5 The author at a talk in Beijing in 2015, the year in which China started to pass Germany as the world's leader in PV.

- For further stimulation, the **Green Electricity Certificates (GEC)** had been already adopted in 2017. The purpose was to develop a market-based mechanism for renewable electricity allowing businesses and individuals to buy this electricity voluntarily. With the GEC, companies could claim the environmental benefits associated with renewable electricity generation. By September 2017, the GEC trading platform had issued 8 million certificates corresponding to 8 billion kWh of on-grid solar and wind electricity.
- Coinciding with the end of the FIT support was adopted in 2019 the **new Provincial Renewable Portfolio Standards (RPS),** a quota obligation on power supply companies and consumers to sell or buy a proportion of their electricity from renewable sources. Quotas for renewable energies' implementation were imposed on all the provinces.

- **Eventually solar energy and wind energy were made subsidy-free in 2020. China removed subsidies for new solar PV projects from the central government budget in 2021.** Now solar technologies had become close to, if not below, the cost of new coal-fired power in most regions, they said. The advances in cost reductions have facilitated **grid-parity**. **Those days' benchmark prices for coal-fired electricity generation allowed, it was estimated, solar projects to compete with internal return of some 8%. That became the criteria to make PV competitive on the markets.**

The new regulations in those years did not leave practical installation rates unaffected. From 44 GW installed in 2018, its rate dipped to only 30 GW in 2019 when the bottom was reached. The following year, installation rates in China started moving upwards again, eventually at a breathtaking speed, now in a **liberalized market**.

A particular breakthrough was noticed for the year 2023, with a market explosion never seen before. China deployed 216 GW of new PV in the country that year, more than it had deployed in the previous 4 years combined, and more than the rest of the world combined. (While China deployed new 216 GW_{ac} or 260 GW_{dc}, the rest of the world installed only 180 GW_{dc} that year).

216 GW_{ac} of PV power on the grids, that is 260 GW_{dc} of module power, was, too, more than all the panels installed to this date in the United States.

Following the National Energy Administration (NEA), China invested 670 billion yuan, or $94 billion, in the deployment of solar power in the country in 2023.

Also following NEA, China employed 5.1 million people in solar PV in 2023, 70% of global PV employment in the field. PV is by now a relevant sector of the Chinese economy.

China's PV surpassed in 2023 again the installation rate of wind power in the country, as it did since the previous year already. Indeed in 2023 China installed altogether:

- 216 GW_{ac} of **PV**
- 39 GW of new wind power

- 33 GW of **thermal power**
- 7 GW of new **hydro power**, 6 GW of which pumped hydro storage (in 2024 was inaugurated the **Fengning** Pumped Storage Power Station of 3.6 GW, the largest of its kinds in the world)
- no new **nuclear power** plant put in operation—so much to the revival of nuclear power in China.

At the year-end 2023, China had a total of 609 GW_{ac} of PV power installed and grid connected generating 583 TWh of electricity; some 1000 kWh/kW for the year or 2.8 kWh/kW and per day.

The capacity factor for PV in China in 2023 was 14.7%. It was lower than that in the United States, for instance, as China had a large share of distributed PV in the country. The 14.7% was an average; like in all countries and regions it varies with the solar resources on each site, it varies from month to month and is lowest in wintertime.

The lower capacity factor in China is in part due to the big share of distributed PV in the country compared to utility-scale. **Following NEA, some 96 GW of all PV installed in 2023 was distributed and 120 GW utility-scale.** China's solar resource is less than optimal in the coastal population centers in the South and East, where much of the distributed capacity is deployed.

The leading province for distributed PV built up in 2023 were the coastal Shandong with some 40 GW of new capacity, followed by neighboring Henan, a bit more inland, a particularly populous province...

China's best solar resource comes more inland, in the North, with Inner Mongolia, and the Western provinces. In particular, the big deserts there offer much solar radiation. Those are more suitable for the deployment of large utility-scale PV plants and are promising for higher capacity factors.

In 2024, the total PV capacity of 886 GW reached in China, 277 GW installed in the year, +45% y/y (figures from NEA). Electricity produced from this power grew as well by 28% year on year.

It grew to over 1000 GW during May 2025, or 1 TW. Following NEA, an increase of 56.9% since a year earlier, some

200 GW was installed in the first 5 months of 2025, a new explosion in the home market. An improved capacity factor is expected with more capacity set up in sunny inland provinces.

The bad side of this very fast growth is the overcapacity in the industry that comes with it: raising in 2025 a $1 billion deficit. The government reacted by abolishing since June 25 the guaranteed tariffs, which have been replaced by local tendering.

China PV Industry Association (CPIA)

In March 2024 CPIA published its latest roadmap. It came with aggressive objectives, an exponential manufacturing growth across the industrial supply chain:

- ramping up more than 750 GW of modules in 2024, a 50% growth over the 500 GW delivered the year before
- producing 820 GW of solar cells, from 545 GW of cells in 2023
- expanding the production of silicon wafers to an equivalent of 935 GW
- producing 2.1 million ton of polysilicon in 2024

Later that year the Association came up with more warning comments. There are problems in the bush. **A need to give more attention to curb below-cost pricing practices and avoid unhealthy competition was recognized. Indeed, in China prices have fallen below cash production costs for poly, wafers, cells, and modules.** Despite tripling capacity, industry profits went down by 70%, it is said. Multiple small domestic manufacturers have already undergone restructuring.

Electricity storage

A new highlight in 2023 was China's explosive growth of PV electricity storage, too. The installation of 20 GW_{ac}/48 GWh new storage capacity was 10 times the installation rate of 2020. It brought the total at year-end 2023 to 31 GW_{ac}. Another record of battery storage installation was achieved a year later with 74 GW/168 GWh, a y/y increase of 130%. 76 billion yuan have been invested in storage. A new trend is that storage power projects are becoming more centralized and large scale.

Storage comes also a lot with residential PV where lithium batteries dominate the market. Storage time exceeds on an average just 2 hours. In 2024, it was mandatory in 26 provinces. PV projects there have to include storage of 12% of the PV rated power. 36 GW of the smaller BESS were added in 2024.

Battery storage in combination with PV is now overtaking the battery market for electric vehicles.

Modernizing the grid infrastructure

In 2024 China invested 608 billion yuan ($83 billion) in power grid construction. Investment in DC projects increased by 227%, most of them long-distance ultra-high-voltage (UHV) transmission lines to connect the PV from Western China to the population centers in the East. 3 of them were put into operation that year. For the same purpose, approximately the same amount is invested in 2025.

Here more about distributed PV capacity in China

The lead of distributed PV was triggered by a countrywide promotion by NEA for the Chinese government. In 2021 NEA rolled out the "Whole County PV Program." The program encouraged counties to build rooftop PV that covers at least 50% of government buildings, 40% of public buildings, schools and hospitals, 30% of commercial buildings and 20% of rural homes. (*It was mentioned in one of my previous books*).

Over 676 counties in 31 provinces have signed up, most of them in Eastern China. Local governments hoped to lower the cost of electricity and revitalize rural China.

And it worked. In 2021, the centralized PV capacity deployment of 24 GW was lower than the distributed one of 29 GW, including 21.5 GW for the residential part.

In 2022, distributed PV installations led the market again with 51 GW over 36 GW only for utility-scale power additions. Among the distributed PV deployment commercial and industrial C&I came 26 GW and the residential part 25 GW. The further strength of decentralized deployment benefited from falling equipment prices.

In 2023, total deployment, as we saw before, took such a speed that the distributed part could hardly follow, it went y/y from 51 GW to 96 GW as centralized deployment went from 36 GW to 120 GW, an increase of 300%. The policy to deploy PV further inland was bearing its fruits.

However, in 2024, distributed PV surpassed centralized again, if only a little. In H1 2024, from the total of 102 GW newly installed PV, centralized accounted for a mere 48%. The residential sector declined from 50 GW in 2023 to 16 GW in the half year H1, but the C&I exploded to 37 GW in H1, as much as the 37 GW installed in C&I for the whole year of 2023.

The residential sector started suffering from the limitations of the local grids that got overloaded. The market of battery storage in association with decentralized PV started to take off.

Centralized PV benefited from larger deployment further inland and in the North, such as Inner Mongolia.

Here is more on centralized utility-scale PV deployment in China

The central government came up in 2021 with the idea of initiating "mega wind and solar bases." Three waves of them were planned. The first started in 2021 covered 19 provinces and had a capacity of 97 GW. Later these "mega bases" disappeared from further discussion, but the start of the "first wave" was certainly a big incentive for going ahead, namely with PV on a large scale.

Whatever, for large utility-scale solar PV deployment the country went to the "Far West."

The northern and northwestern provinces got the lion's share of the new utility-scale plants. Despite their remoteness, those are attractive as they come with a sunny resource and land is cheap. And the transfer of electricity to the populated provinces in the East high-voltage transmission is on the cards and is under construction.

In 2023, the leader of new centralized PV capacity was the autonomous republic of Xinjiang, in central Asia. 38 GW of PV were installed in this huge area that is sparsely populated and known for its dry steppes and big deserts. **Near the region's capital,**

Urumqi, was inaugurated by mid-2024 the world's biggest PV power plant, a 5 GW giant on 200,000 ha of land to generate 6 TWh of electricity per year.

A new, even bigger 8 GW plant is in the planning stage at Ordos to cover an area larger than that of Singapore. Ordos lies in the Kubuqi desert in Inner Mongolia, where in 2023 altogether 23 GW of PV were set up. PV is becoming popular in a county that hitherto was known to be a coal county—not a small progress. But that's not the end of the story:

Ordos is planning a Solar Great Wall. Launched in 2024, it should cover an area of 400 km × 5 km. When finished in 2030, the 100 GW project should generate 180 TWh of electricity. This clean electricity could be delivered to Beijing. Another purpose of the new Wall is to stop desertification in the area.

Just South of Inner Mongolia in Northern China lies Shanxi with 25 GW of new utility-scale PV deployed in 2023.

Not to forget the very sunny high plateau of Tibet. In the large Western province of Qinghai, which includes a big part of this plateau, even 28 GW were newly set up in 2023.

Figure 2.6 The author ballooning in China, 2008.

2.2.1.1.1.2 *China's PV exports*

China's solar modules are very competitive on the international markets. They are exported worldwide to some 200 countries.

In 2023, China exported 227 GW of PV modules; the corresponding market value amounted to $47.6 billion.

In 2024, exports of PV modules increased by some 15% to over 240 GW, but its export value dropped by some 35% to less than $40 billion. Export volumes were stimulated among other effects by the steep cut in module prices on the international spot markets as mentioned previously, some 13 cents/Watt compared to 18 cents/Watt in 2023.

Eighty-nine percent of the modules installed in *Europe* were imported from China. But in H1 2024, only 54 GW of the 132 GW of China's PV module exports went to Europe, a 13% drop from the previous year. In 2023, Europe had taken some 50% of China's module exports. Module prices in Europe stayed low at some 10 cents/Watt and were not affected by higher shipping costs by the Red Sea crisis. The reasons for the falling imports may be linked to the high inventories from previous years and the slowdown in demand.

The Asia Pacific market accounted in H1 2024 for 40 GW, a 120% increase over the preceding year. More than half of these exports went to Pakistan and India. Australia's imports increased by some 50%, while those of Japan and of most other countries in the region went a bit down.

America's imports increased slightly to 17 GW of module imports. By far, most of these imports went to Brazil. **Imports to the United States are negligible.**

The Middle East imported 15.5 GW, a 163% increase. Saudi Arabia was the region's largest importer stimulated inter alia by the Saudi "Vision 2030" program.

Africa imported 5 GW of modules, an increasing trend. South Africa was the top importer.

Top destinations of module exports in 2024 were the Netherlands, Pakistan, Brazil, Saudi Arabia, India, Spain and 200 other countries around the globe.

- **In H1, the Netherlands imported 23.5 GW of PV.**
- **Pakistan imported 12.2 GW.**
- **Brazil took 10.5 GW in H1.**
- **India got 8.3 GW.**
- **The Middle East got altogether 15.5 GW: Saudi Arabia 7.6 GW, UAE 2 GW, Oman 1.4 GW, Qatar 0.8 GW, and others.**

2.2.1.1.1.3 *China's Electrochemical batteries export*

In early 2024, Chinese exports increased 50% in this new sector.

For instance, **BYD** has sold 75 GWh for 350 projects in 110 countries. In February 2015, **BYD** signed an agreement in Saudi Arabia for 12 GWh for their Vision 2030 project. BESS will be grid connected on five different sites. The batteries will be the new model ML Cube-T.

Figure 2.7 Last book presentation with Qin Haiyan in Zhangjiakou in November 2019, just when COVID-19 was about to break out.

2.2.1.1.2 *PV in the United States*

America was and is a great solar nation. And has nothing to do with China, the solar giant. That's the way it is today. (**A lot of details can be found in the working paper of the Harvard**

Business School: "Power from Sunshine, a Business History of Solar Energy," 2012, G. Jones and L. Bouamane).

Solar PV added more new capacity to the grid in 2024 than any other technology in the last 2 decades, found the US Solar Association SEIA.

New PV stood in the United States for 66% of the new generating capacity for the year 2024. New wind power came in for 13%, Natural gas for 4%, and a little 2% for nuclear. All the new capacity added up to 62.8 GW.

Following SEIA, 50 GW_{dc} of PV were newly installed in the United States in 2024, bringing the total to 236 GW of capacity. It was up 21% from installations in 2023. Electricity storage came in 2024 also with a new record of 12.3 GW, an increase of 33% y/y; by 2029 81 GW more are expected. Solar PV together with storage made 84% of the new capacity. PV and storage were now more affordable than any other technology in the United States.

Forecasts see continued growth: by 2035, a US PV capacity of some 739 GW is expected, compared with 236 GW in 2024.

According to the consultancy Ember: Solar generation increased by 64 TWh in 2024. Wind power and PV together reached 17% of total electricity generation in the United States in 2024, surpassing coal for the first time.

PV continued to be the fastest-growing source of electricity surpassing hydro generation for the first time that year.

41.4 GW_{dc} are utility-scale solar in 2024; residential was 4.7 GW, in decline. More than half of the new utility-scale capacity comes in three sunny states: Texas (35%), California (10%), and Florida (6%).

In its outlook from early 2025, the US EIA anticipates for 2025 an increase in solar electricity generation by 34%, and 18% more a year later. Eventually PV may provide 27% of US electricity supply by 2026.

The capacity factor for PV in the United States was on average 20% in 2023. It varies by a few percentage points from year to year. It is highest in the South West of the United States, is higher for utility-scale than rooftop, is lowest between November and January, and no surprise, can exceed 35% in summer—in some states.

Most of the PV installed was with silicon solar panels. Some were imported from Vietnam, Malaysia, Thailand, and Cambodia. But for the first time, most of the 50 GW of modules installed, 42 GW, had been produced in the United States; in 2023 it had only been 14 GW of domestic production. As mentioned earlier, in April 2025, Trump brought all PV imports from South East Asia with 3000% import duties to a halt.

Silicon solar cell manufacturing resumed a bit in the USA, too.

On the other hand, the United States' industry does not export any PV.

Module prices in the United States are some 33 cents/Watt. They are quite higher than those on the international spot markets as the United States has protected itself with import restrictions. First Solar, the main domestic producer of modules, had in 2024 a gross margin of $2 billion. No problem of profitability like some other module producers.

SunPower, one of the oldest and most prestigious PV firms in the United States, went down the drain on Wall Street by mid-2024 and was bought up by others (It was owned since 2011 for a third by the French TotalEnergies). It had specialized in rooftop installations, a market, the residential PV market, that, as mentioned here above, broke virtually down in 2024.

Likewise the Swiss company **Meyer Burger** has serious restructuring problems for its silicon module plant in Goodyear AZ. Eventually, by June 2025, the company gave up. All its workers in the United States and in the German production sites got laid off.

In 2024, residential PV plus storage systems in California had a median price of $5.518/kW$_{dc}$ (NREL Solar Industry Update). Given the dramatic decline in solar module prices, the heart of the system, such prices are rather on the high side.

ACORE, the American Council on Renewable Energy (once initiated by H. Scheer and Mike Eckhart), gave some further information in a paper it published in early 2025, with S&P commodity insights. From it, the following could be concluded:

- Solar PV and wind power–generated electricity are the cheapest in the United States. They come at some 5 cents/kWh; in comparison, new nuclear exceeds 28 cents/kWh.

- In 2024, 91% of new utility-scale capacity deployed in 2024 was PV, wind power and storage: PV +25 GW, wind power + 5.5 GW, (nuclear + 1 GW).
- In 2025 with $670 billion of investment, clean energy comes for the first time upstream oil and gas: for PV some $370 billion, wind power $150 billion, offshore wind power $90 billion, and storage $40 billion.
- With 63 GW PV announcements for 2025, it will account for 50% of new clean power capacity additions, wind power will take 13% and battery storage 35%.
- BESS will surpass for the first time pumped hydro capacity.
- 82 GW were invested in clean energy PPA over the last 5 years.
- 6 million students in the United States learn in schools equipped with PV.
- China adds twice as many GW of clean energy per dollar than the United States.

As far as employment was concerned, in 2023 some 300,000 jobs were associated with PV, mostly in the construction sector.

Comment: *When President Trump came to power in 2025, nobody had expected him to install solar panels on the White House as President Carter did long ago. Cynics would say that before doing that he might rather put an oil-drilling rig in the Rose garden of the White House.*

The reality looks so far different. There is no reason to believe that the new president would do much harm to solar energy development in his country, leaving alone that Elon Musk, his Minister is rather known as a supporter of solar PV and storage; that's part of his business.

It is true that Trump's United States is isolationist and seems to be living nowadays on a kind of separate solar planet for its energy matters. Indeed, all bridges are cut by now already with China, the leader of solar power for the rest of the world, at least no PV imports from China for the whole value chain and no cooperation, as is the case for 200 other countries.

Figure 2.8 The author, on the left, at an oil-drilling field in Azerbaijan, one of the world's oldest.

There is no risk of a potential dispute on patents. Today's preferred TOPCon technology for silicon solar cells and the CdTe thin-film technology, too, both have been "invented" in Germany.

Hence, the United States is already great in solar matters today, it is not waiting for a new solar hero. It was the Biden Administration that did it during its own term. Apart from trade taxes to create a Great Wall against much solar energy import, the Biden Administration had established IRA, the Inflation Reduction Act. (see the book *Solar Euphoria*, 2023, p. 30.) *It is undeniable that IRA stimulated considerably the deployment of solar energy and the other Renewables in the country.*

On the other hand, it did its part to isolate the US markets, namely with SEMA, Solar Energy Manufacturing for America. Since IRA became effective in 2022 more than 85 GW of manufacturing capacity has been added across the supply chain. The access it provides to tax credits for domestic manufacturing across the value chain changed the picture for America in the world.

*Then eventually Trump came with **BBB**. This does not mean Big Bullshit Bill, even though many people think it should. Its official name is Big Beautiful Bill. After approval by Congress by a low margin it was signed into law on July 4, 2025. One who does not like it in particular is Elon Musk. His big friendship with the president is broken and he wants now to create his own political party.*

The BBB is not favorable to Renewable Energies; any cooperation with foreign countries, in particular, is heavily punished. As could have been expected, the Law provides support in the United States for fossil energies and nuclear. Fortunately, Trump cannot impose his views and Laws globally!

2.2.1.1.3 PV in Europe

Following the Info Link Database, Europe saw for 2024 new PV installations of 85 GW or more. It was the first year when solar PV was leading electricity growth in Europe in terms of both capacity and generation. Consequently this led with wind power growth to a decline in the use of coal and other fossil fuels. The carbon price stabilized in the range of some €70/tCO_2 (see more on the EU ETS, the GHG Emission Trading System in the Section 2.1, "Introduction").

Of the up to 85 GW of new PV capacity deployed in 2024 in Europe, the leading countries with their share of the total were thc following:

- Germany 19%
- Poland 8%
- Italy 8%
- Netherlands 7%
- Spain 7%
- France 5%
- Portugal 5%
- UK 4%
- Greece 4%
- Belgium 3%
- The remaining countries together 30%

Solar electricity generation rose by 20% in 2024. Generation from PV was a quarter of total renewable electricity generation that year in Europe. It was the same as the share of wind power generation. Hydro had a share of 35%, Bio-power 6%, and wind power offshore 4%.

All renewables covered some 50% of demand already in 2024, an increase of more than 10% over 2023. Fossil fuels and nuclear each covered a quarter of electricity demand,

The 27 EU Member countries were adding 65.5 GW PV in 2024, a 4% increase over 2023, says Ember.

In 2024, the EU set a new record for **renewable electricity supply**; for the first time, it was 47.4% of the total, almost 50%. Next to PV with 11.1%—that was more than the 9.8% from coal—wind power provided 17.4%, hydro 13.2%, Bio-energy 5.5% and others 0.3%. Thanks to France, Nuclear generated 23.7% of the EU's total electricity supply.

€56 billion was invested in new PV in 2024, almost all from private sources.

Forty-two percent of the new installations were utility-scale, 39% were C&I, and at 12.8%, only 20% was residential. In 2023, residential came 33%, a percentage as big as those of utility-scale and C&I that year.

The EU's total solar capacity increased to 338 GW, (725 W/ capita). It is an enormous increase over just 120 GW the EU had installed 5 years earlier in 2019.

By 2030, the EU target for the deployment of PV comes 750 GW_{dc} or 600 GW_{ac} following REPowerEU; but current growth suggests a higher capacity, maybe over some 800 GW.

Consequently, the EU comes in 2024 second in terms of the world's total solar deployment after China and before the United States that has 236 GW installed.

The 338 GW were generating 300 TWh of electricity. The average capacity factor for Spain was over 20%, more than that of the United States. For Germany, it was lower than 10%, an effect of degraded weather over the past years, with more rain and less solar radiation. But weather caprices make 2025 sunnier again to the benefit of electricity harvest from PV installations.

Virtually all panels deployed in the EU were crystalline silicon modules; 89% of them were imported from China.

At year-end 2024, prices for modules with TOPCon silicon modules had reached a bottom of less than 8 cents/W in Europe.

The price for a complete PV rooftop system can be as low as €1500 /kW. That is less than a third of the price of such systems in California.

PV provided some 830,000 jobs in Europe in 2024.

The hours of negative wholesale prices of electricity increased in 2024. Obviously there is a need for better interconnectivity, demand-side management and storage. It is also an incentive for more self-consumption.

Residential storage is developing in the EU at great speed. 14 GWh of home battery storage are installed here in 2024, 10 times it was 4 years ago. The PV market for stationary battery storage stayed well behind the one for EV but it was progressing by 61% in the year. Prices decreased by 43% y/y. A battery for 2 hours of storage cost $115/kWh in April 2024.

In the C&I sector, one counted 0.7 GWh installed in 2024.

Later in May 2025, SolarPower Europe came with new figures for 2024. They found 21.9 GWh of BESS newly installed in Europe in the year, bringing the total to 61.1 GWh they found.

Half of the batteries in Europe, or some 500,000 of them in 2024 were installed in Germany.

And what about utility-scale batteries for PV? It's a brand-new issue in Germany; at year-end 2024, only 1.5 GW stood on the grids or 1.8 GWh. But, as they say, there is a boom of interest for the "super batteries," a tsunami of storage for PV is on the horizon: indeed an interest of astronomical scale for 161 GW of new batteries for PV on the grids there is. Even if only half of them were built it would mean some 230 GWh of stored PV electricity. Leaving alone that development work is underway to augment the storage duration per GW. *So, PV has transformed from an intermittent electricity supplier to one of 100% security like coal, oil or gas. Another energy revolution.*

Currently, PV in Europe is primarily deployed on rooftops, for C&I and residential applications as it were, and marginally as agrivoltaics or floating PV on water. What about area constraints for extra-large-scale PV deployment when it is ground-mounted? Here comes other good news:

In July 2024, an analysis by the European Environmental Bureau (EEB) revealed that Europe has sufficient land to expand solar and wind energy without compromising food production or nature. The study showed that only half of the land deemed suitable for renewable energy—excluding natural reserves and high-value agricultural areas—is needed to decarbonize the EU by 2040.

Only 2.2% of the EU's total land will be required for future solar and wind projects to allow the EU to phase out both fossil fuels and nuclear power to reach climate neutrality by 2040. 5.2% of the EU land can be considered suitable for solar and wind power development, based on strict agricultural, environmental and technical criteria for hosting onshore wind and solar projects. Rooftop is not enough, they say. Urban and industrial areas alone can't meet the solar capacity needs. But there is plenty of degraded agricultural land available to expand PV without disrupting rural economies. It can be done in synergy with food production and soil-health restoration.

Area needs of PV and Golf courses: In February 2025, the *Journal on Environmental Research Communications* published an article by J. Weinand from the Jülich Research Centre in Germany. It found that German Golf courses—there are a thousand of them in the country—need more land area than PV plants; a quarter more actually. The author took for PV an area need of 0.01 km^2/MW. For wind power, the relation is a bit less favorable when adopting 0.12 km^2/MW for it.

The situation is similar in the UK, Canada or the USA, where area needs of PV are currently lower than those occupied by Golf courses. For a 100% supply of the United States' energy needs with PV and wind power, the author calculates an area occupation need of 38,000 km^2, equivalent of the total area of Switzerland—5 times the area of all US Golf courses today. Obviously the author ignored here the tremendous opportunities for PV rooftop applications, as well as offshore wind power.

Negative electricity prices in Europe in 2025

(*see also Chapter 4, by Jürgen Sachau*)

Solar PV and wind power capacities have doubled just in the last five years. In particular, PV in Europe has seen an exponential growth over the years. And everybody benefits from it, everybody, not only the investors in PV. So electricity prices have declined in 2025 again on the grids, augmenting the trend of the previous years.

Following Ember, the average monthly electricity day-ahead prices declined everywhere in Europe, for instance, in May 2025, in Germany by 15% down to 6.6 cents/kWh, in Spain by 43% down to 1.5 cents/kWh, and so on.

And even better, on certain summer days, electricity generation by the PV installed is high while demand may be low for holiday seasons, Sundays, or other occasions when demand on the grid from industrial and commercial customers is lower; on such situations, the grid operator pays a bonus to buyers for getting rid of the excess of electricity they got to handle: *the negative electricity prices.*

According to Enervis, the number of hours with negative prices rose across Europe in the first half of 2025 with its many sunshine hours. The highest number was noted in Spain, 404 hours, in Germany it was 389 hours. (For the whole year of 2024, it had been 457 hours, and even less in 2023: 301 hours.) A paradise situation: people can charge their EVs, drive heat pumps, etc., for free and also earn money by doing so.

The situation will not change too much in coming years before more storage capacity was massively deployed—see the following paragraphs. More electrification of demand and better grid interconnections and distribution will also help.

EU solar energy strategies

The EU has been the major driver behind Europe's solar market development. In 2019, the EU Commission introduced the European Green Deal pledging to achieve net-zero emissions by 2050 and a climate neutrality goal.

As part of the **REPowerEU Plan** (see our book *Solar Euphoria*), the EU Commission adopted later in May 2022 an EU solar energy strategy which identifies remaining barriers, challenges, and design initiatives to accelerate PV deployment. It aimed to achieve over 320 GW PV in 2025 (was already done actually in 2024) and almost 600 GW by 2030.

A set of new initiatives are outlined in a revised Renewable Energy Directive from 2023 (EU/2023/2413). The EU solar energy strategy launched three initiatives, as follows:

- **European solar rooftop initiative.** A proposal to gradually introduce a "solar ready" obligation for new buildings. They were actually included in the "Energy Performance of Buildings Directive." For existing public buildings, solar will have to be gradually installed, starting in 2027. It's not aggressive as it adds "were technically, economically, and functionally feasible."
- **EU large-scale skills partnership.** Launched in 2023, it aims to promote the development of a skilled workforce for PV.
- **EU PV Industry Alliance.** It was launched in December 2022 as a forum of stakeholders focused on ensuring investment opportunities and supporting the domestic PV industry. (*It is an ambitious enterprise as Europe's domestic PV industry, apart from a few exceptions, is in jeopardy).*

The European Solar Charter. In the margins of an informal Energy Council meeting on April 15, 2024, energy ministers from 23 EU countries and the competent Commissioner signed a European Solar Charter. It sets out a series of voluntary actions in support of PV, in particular the support of panel manufacturing in Europe.

As part of the EU's **"Net-Zero Industry Act"** to boost manufacturing of net-zero technologies in Europe, **the European Solar Academy** was launched in June 2024. Its purpose is to promote working skills, aiming to train 100,000 workers over 3 years.

And there is the EU trading system ETS for GHG emissions from fossil sources since 2005 and ETS2 from 2027 (see the introduction of Chapter 2 of the book). It weakens the renewable energies' fossil energy competitors giving an advantage to PV, wind power and all other clean renewables.

Additional PV power capacity installations in the 10 leading EU markets in 2024

- Germany 16.1 GW
- Spain 9.3 GW
- Italy 6.4 GW
- France 4.7 GW
- Poland 4.2 GW
- Netherland 3 GW
- Greece 2.9 GW
- Austria, Hungary, and Portugal (Belgium is no longer among the top 10 EU country markets in 2024)

Since 2025, PV and wind power have provided more electricity in the EU than all fossil fuels combined.

2.2.1.1.3.1 *The lead of Germany*

Germany is a solar energy pioneer. In 2024 again it took the lead of PV deployment in Europe.

In 2024, a great 62.7% of the electricity produced in the country was renewable, an increase of 9% y/y while the share from the fossil sources kept declining. Since all nuclear plants had been stopped, the country generates no nuclear power anymore. Among the renewable electricity produced, 33% originated from wind power, 17% from PV, 6% from Bio-power, and 5% from hydro.

In 2024, the country had in total some 98 GW of PV capacity installed, some 16 GW on top of the 82 GW in 2023; an increase of 10%. 72 TWh of electricity were produced.

Deployment of PV in Germany keeps accelerating; in April 2025 already 104 GW stands in the country. That's good for European conditions, but only 10% of the PV that was installed in China!

In 2024, 1 million new PV installations were bringing the total in the country to 5 million. 2/3 of deployment in 2024 came on rooftops. 70% of them were associated with a battery, mostly LFP batteries of lithium, iron, phosphate. Germany is famous for its plug-in PV systems. The country had some 800,000 of them installed at year-end 2024 as **balcony PV systems**. They come at low power of 1 or a few kW, including the inverter and, as an option, with a battery. They are sold tax-free.

For larger PV systems, be they rooftop or ground-mounted, the state bank KfW can provide a credit of some 5% interest rate, limited to 2 years.

Ground-mounted PV systems were still running in 2024 through public tendering. The average tendering price in 2023 was some 5 cents/kWh.

In a liberalized market, **solar PPAs** are developing fast on a GW scale. An example is the PPA between RWE and the Salzgitter Group, the latter acquiring the electricity generated from the 180 MW Boitzenburger Land in Brandenburg. That one was completed in September 2023. It has 320,000 PV modules and was built in only 11 months.

Another example for commercial operation is the **650 MW** PPA project "Move On Energy" at Witnitz in Saxony near Leipzig. It was completed in July 2024 and is **Europe's largest** PV project deployed on 500 ha of a former brown coal mining ground.

As mentioned above, Germany altogether had in 2024 some 5 million solar systems grid connected and running.

The kWh injected into the grid is paid at 12.7 cent/kWh for a complete injection of all current produced and 8 cents/ kWh when only a share is grid-injected. Payments are lower for larger systems over 10 kW. They have decreased in small steps over the years.

In the previous section on PV in Europe, we had mentioned the new interest for battery storage on a very large scale for deployment on the electric networks. **Germany is the main European country going forward for this new interest in storage. In this respect, it is worth mentioning a new initiative by the Grid Operating Agency to lower in future the connection cost to the grid that can be considerable at present.**

Today the grid operator does not demand a connection fee for all systems feeding electricity into the network, **such as PV**, wind power, etc. Those who have to pay for being connected are the users such as factories, also electrolyzers for clean hydrogen generation, grid-connected **storage plants** for wind power, **PV,** and others. The fees that can amount to many millions of Euros could be reduced in future by some 90%.

2.2.1.1.3.2 *Other frontrunners of the EU's PV deployment*

2.2.1.1.3.2.1 *Poland*

Poland was in 2023 the second-largest solar market in Europe. Poland's PV was catching up: five years earlier, it was only fourth place in Europe. In 2024, its market fell back again to the fifth position among the EU's PV markets.

Reuters found out that Poland and Hungary are the most important drivers of solar electricity generation in Eastern Europe; in the first 7 months of 2024, Poland produced 11.3 TWh of it, 1/3 more y/y.

In Austria, Bulgaria, Hungary, Romania and Poland utility-scale solar output jumped by 55% in H1 2024, a much faster growth rate than that of the rest of Europe.

Poland had deployed in total 17 GW by year-end 2023. In that year the country had 4.6 GW of PV newly installed.

In 2024, the rate of installations decreased y/y, from 4.6 GW to 4.2 GW and a total of 21.2 GW was reached; 43% of the new PV were micro-installations, the rest came mostly in plants over 1 MW capacity.

The cumulative PV capacity comes as the lion's share of all renewables in Poland. At year-end 2023, the country had next to PV, 9.4 GW of wind power installed, 0.95 GW of hydro, 1 GW of solid Biomass and 0.3 GW of Biogas.

The country still relies heavily on fossil fuels for its energy supply and is a strong emitter of CO_2. But there is hope: in 2024 already 27% of Poland's power capacity was PV and 7% of all electricity generation was solar.

The country is facing grid overload issues in off-peak hours. PV plants had for the first time to be disconnected from the grids.

In August 2024, Poland launched round 6 of a prosumer-scheme for micro PV installations funded with $104 million. The scheme is financed by the EU program FEnIKS that supports the country in its fight against climate change and for a cleaner environment. PV systems applying for support must be associated with a battery.

2.2.1.1.3.2.2 *Italy*

In 2024 Italy ranked third again for PV installations in the EU. With 6.8 GW newly deployed in Italy or +30% y/y, the country reached 37 GW of PV capacity year-end 2024. 1.8 million individual PV plants were installed, most of them in the North. By mid-2025, the total PV capacity reached even 40 GW with over 2 million plants in operation.

Italy's power capacity of all renewables reached already 60% of total capacity in 2024. With 46% of the renewables' share, PV power came first, followed by water power (29%) and wind power (19%).

As expected, water power was in turn the frontrunner in terms of supply and consumption of electricity; it supplied about half of the 41% of the renewable electricity share. PV-generated current provided 10% of the total supply, followed by wind at 8%. Natural gas and a small part of coal contributed together 35% to the supply. The rest of some 19% was imported.

TERNA, the grid operator, reported that Italy has increased by 41% year on year its PV power capacity. The growth is mainly due to larger utility-scale plants in the center and South of Italy while the number of new PV plants decreased. Ninety percent of the new PV plants installed were residential ones. But only in terms of numbers, in total power, they were not more than a quarter of all new installations.

The loss of enthusiasm for the domestic segment may be due to this year's reduction of the subsidy rate by the super-bonus scheme for distributed generation from 110% to 70%. On the other hand, the government launched in March 2024 the National Recovery and Resilience Plan 2 offering subsidies for developers that use "homemade" solar modules. The EU Commission, through Italy's Green Deal Industrial Plan, makes incentives available for local PV manufacturing.

Unlike the residential sector, the C&I segment of the PV market increased in 2024 to over 22% of total solar power capacity additions.

The utility-scale sector for plants with over 1 MW capacity represents over 50% of the new PV power installed.

Battery storage also saw a significant uptick in 2024. Data from TERNA suggest that Italy installed in H1 127,000 storage systems with a total power capacity of 1.05 GW or 2.6 GWh, bringing the total to 650,000 systems in operation. Most capacity units are in the range below 50 kWh for residential use. Their number was in decline in H1 2024 reflecting the end of the Super-bonus and the "credit transfer block."

There are only a few larger storage capacity systems in operation for standalone PV systems, the biggest in Udine.

For the future: At last Italy has a National Integrated Energy and Climate Plan (PNIEC), approved by the EU Commission after harsh negotiation. It is the main roadmap for Italy's decarbonization. Towards the year 2030, in total 131 GW of renewable capacity is planned as follows:

- 79.2 GW of PV
- 28.1 GW of wind power
- 19.4 GW of water power
- 3.2 GW of bio-power
- 1 GW of Geothermal Power. Italy has a long tradition in Geo-power; see the Lardarello plant from 1913; it was the world's first of its kind.

2.2.1.1.3.2.3 *The Netherlands*

With over 1.5 kW per head, the Netherlands ranked first in Europe for the solar power installed per capita in 2024. It was double what it was 3 years earlier.

PV stood for 17% of the Netherland's net electricity production in 2023.

In 2023 the country had added 4.3 GW to its PV capacity bringing it to 23.9 GW in total. In 2024 new deployment of 3 GW of capacity brought the total to 26.9 GW. Forty-one percent of it was residential, and the rest was for C&I applications. Of the more than 2.8 million PV installations, the largest share of 2.5 million systems were installed rooftop. 2.6 million PV panels were installed on homes.

One in three homes has PV on the roof. Currently residential PV owners use net metering; all goes through the same counter, irrespective of whether they buy electricity from the grid or inject it into it. It's simple and very favorable for the user. But it has now been decided to stop it by 2027, and one has to switch to a dual flow counter. There is a new EU directive, too, that imposed its removal from 2024 onward in the EU countries.

What is good for the user may be bad for the state budget: with net metering, no VAT is collected for the current that otherwise would be bought from the net were it not provided by the PV on rooftop. So the Dutch taxmen lost altogether some €3 billion.

Next to net metering, the Netherlands also uses an auction system, the subsidy by the SDE++ Decree, formerly approved by the EU. A new round of it was started in September 2024 with €11.5 billion. It cannot be used for self-consumption PV plants.

PV installations lost a bit steam in **2024** compared to the previous year, as is the case everywhere in Europe. Only some 300,000 new PV plants with 3 GW of power were installed in the year.

To fight against increasing grid congestion with the breathtaking success of PV was an incentive to more associating batteries with new PV deployment. By 2030 some 10 TWh of new storage are now planned.

2.1.1.1.3.2.4 *Spain*

In January 2025, Spain had in total 32 GW of PV grid connected, just a tiny bit more than wind power. So, in early 2025, one-fourth of Spain's power capacity was PV. In 2024, PV generated 44.5 TWh, 17% of Spain's total electricity supply.

It is interesting to note that PV generates more electricity than hydro, leaving alone the PV for self-consumption, on rooftops as it were.

Year-end 2024, Spain had in total 8.5 GW of PV installed for self-consumption, generating 9.2 TWh of electricity in the year. Most of the power was for industrial application, and only 2.2 GW of the total for residential. In 2024 only 1.4 GW of new PV came additionally for self-consumption.

Spain is well known as a pioneer in PPAs, in particular solar PPAs. In the first 9 months of 2024, it concluded the lion's share of the hundreds of European solar PPAs that amounted in total to 11.4 GW during this period. Two largest European deals were concluded recently in Spain with two pharmaceutical companies at 5 cents/kWh.

On the horizon of 2030, Spain has agreed, too, with the EU an PNIEC action plan.

It foresees by then for the country's power capacity a share of 76 GW of PV + 19 GW PV for self-consumption. Altogether the renewables should by then generate 81% of electricity supply. With an investment of €308 billion also 560,000 new jobs should be created in the clean energy sector; 22.5 GW of storage capacity are planned and 12 GW for electrolyzers for green hydrogen production.

2.2.1.1.3.2.5 *France*

Figure 2.9 The author's good friends in Paris: French solar pioneers Bruno Civel (left) and Alain Liébard (right).

France is the number 1 solar pioneer in Europe, if not in the world. One example is Edmond Becquerel, who invented PV in the 19th century, "in Pre-historical Solar Times." France was also a leader of PV and all renewables in modern times. A milestone was the UNESCO Congress on solar energy that took place in Paris in 1973 with a thousand participants from all over the world. This Congress was an initiative by French pioneers.

France was also a European pioneer in atomic energy deployment and it is more than ever until this date. And both solar PV and nuclear don't fit well together. In our book *Solar Euphoria*, we gave some details about the wide anti-nuclear movement in France over the years in the last century.

But currently France is still a fierce supporter of atomic energy. Not only the young president and his government, but the French people are also supportive of it as well—at least to some extent. A possible reason is that at times of international conflicts and tensions with the risk of power disruptions, French people are happy to benefit from a reliable and secure electricity supply.

The aging of its 56 atomic power plants and the corrosion problems everywhere may be a problem for the engineers of EDF; one can rely on the system works and the competent engineers at EDF. And in 2024 France became even again a big exporter of nuclear electricity to neighboring countries. That's the way it is seen, my own residence is in Paris...

But still: some good news came in early 2025, in a poll. IFOP found out that 51% of the French people currently favor the future development of renewable energies—with PV coming on top. Only 25% give a preference to nuclear power.

In this environment, it is noteworthy that France has kept to this date a small but aggressive solar energy and renewable energies community.

And it has to accommodate further bad news concerning local PV production: end of November 2024, Photowatt, the last industrial hero of PV in France, one with the longest record, did throw the towel.

In 2024, France installed 4.7 GW to reach a total PV power capacity of 25 GW. 92% of the installations were for a power capacity below 9 kW. But those represented only 20% of total new PV power installed. The lion's share of 80% came for utility-scale systems of 500 kW and above.

In total, the PV park installed met some 5% of France's electricity consumption.

France has a Renewable Energy Acceleration Law from 2023. It mandates solar installations on new and renovated buildings over 500 m^2 and parking lots over 1500 m^2. It also encourages PV to be installed on unused land and near roads and railways. For larger systems, tenders are organized.

For the future, a Multi-Year Energy Programme PPE was revised in November 2024. It sets a goal of 48 GW of PV to be installed by 2030.

For 2025, PV supporters in France are less optimistic as the government looks keen to lay big bricks on the road to its further development.

2.2.1.1.3.2.6 *Greece*

In 2024, Greece installed 2.9 GW of new PV bringing the total to 10 GW, says SolarPower Europe. Solar power comes first of all installed capacities for electricity generation in the country. Power from natural gas, wind power, hydro, and lignite comes behind.

In the first seven months of 2024, PV and wind met 45% of electricity demand; together with hydropower, the renewables covered over half of all electricity supply.

2.2.1.1.3.3 *PV in the United Kingdom*

In October 2024 the UK had 17 GW of PV installed, only 1 GW more than a year earlier. That is quite a bit less than Holland's on the other side of the Channel... 5% of UK's electricity supply is solar.

Thirty percent of total PV capacity is residential. 5.2 GW are rooftops. 1.4 million homes have PV rooftop. The first grid-connected PV rooftop in Britain had been a 4 kW one of Sue Roef, the Oxford "Ecohouse" completed in June 1995; it came 5 years after Germany's 1000 roofs programme mentioned above.

Fifty-five percent of PV capacity comes as ground-mounted standalone installations.

Battery energy storage systems (BESS) have been a lot cheaper recently, but the market in Britain did not follow suit and did not develop as expected.

In 2024, there were 1.5 million solar installations in the UK, more than in France. Germany, with a similar climate as the UK, got 5 million.

The UK has 13,000 jobs in solar energy.

The UK has also solar PPAs: the 373 MW Cleve Hill Solar Park with battery storage in Kent is the UK's largest. It will deliver 65% of its production to TESCO covering 10% of their shops' needs, from 2025 onward.

By 2030, it is planned to raise PV capacity to 47 GW. Good planning, we saw above that sunny France has no bigger goal.

2.2.1.1.3.4 *PV in Turkey*

In 2024, Turkey had 19 GW of PV installed. It is more than that installed in the UK—see here above.

Together with the 12 GW of wind power and 32 GW of hydro, the renewables accounted for 60% of total capacity. Nuclear power is under construction but not yet available.

In terms of electricity generation, hydro came first at 20%, wind power–generated current at 10% and PV at 6%.

Current governmental targets for the year 2030 plan 33 GW of PV by then. As deployment keeps accelerating, it is expected that it will be revised upwards soon.

2.2.1.1.4 PV markets in Asia

2.2.1.1.4.1 *Pakistan, India*

Pakistan

PV in Pakistan in 2024 is a success story on its own. It is something extraordinary, satisfactory for many parts of Pakistan's society, so to say bringing wealth, happiness, and new comfort to some of the poorest. Solar PV is an example of the right way of development of an emerging economy.

In 2024, 22 GW of solar PV will have been newly deployed in Pakistan. As we saw earlier in this chapter, it is over 10 times of PV's deployment rate in the UK that year. It is a new triumph: in 2023, "only" 3.5 GW was installed.

Cheap PV module imports from China keep this system going; it works by and large through market forces. It relies on a financing system that benefits all those involved: The Chinese exporters of the ultra-cheap solar modules, their Pakistan importers, and those in the sales and distribution chains until the ultimate beneficiary and user, who is happy to get at last a source of electricity he can rely on. And the whole country benefits from a new rush in clean energy, giving hope to get rid eventually of the terrible pollution problems there—and to help keeping climate change at bay as it suffers these days already from terrible inundations.

What is the reason for this surprising success? We saw previously that since 2020 or so, the Chinese home market started a new explosive growth triggered by new progress in PV technology, making it fully competitive with electricity from coal-fired plants. Today the Chinese market success spills simply over to neighboring Pakistan...

The "conservative" World Economic Forum in an article published end of 2024 says:

> *Pakistan's rapid adoption of solar energy, driven primarily by market forces and with minimal political support, provides valuable lessons for other emerging markets. Declining solar panel prices, coupled with skyrocketing grid electricity tariffs that have increased by 155% over 3 years, are fueling a rush in Renewable energy adoption in Pakistan, with solar power leading the way. The country is now the world's sixth-largest solar market. Industrial, agricultural, and residential sectors have embraced solar, with imported Chinese modules totaling 13 GW in the first half of the year, and forecasts reaching +22 GW by year 2024.*

And a few things helped:

- The return on solar investment is 2 years only.
- There is a generous net metering rate on the grids.
- PPAs are important and there is a lot of C&I involvement for PV uptake.
- Provinces are actively involved:
  - Baluchistan turning PV to agricultural wells;
  - In Sindh, 200,000 panels for low-income households with a 80% government subsidy;
  - In Punjab free solar modules for the poorest.

India

Figure 2.10 The author in India in 2009.

India suffers from heavy atmospheric pollution to such an extent that at times schools must be closed. Heavy flooding in the summer can lead to disasters and even deaths. Solar development to replace the massive use of coal that comes at 55% of energy supply is much supported in the country.

India, one of the world's largest countries, accelerated its solar power growth in 2024. Following the Indian consultancy group JMK, India added 27 GW of PV in 2024, a growth of 4 times y/y. Most of it, over 20 GW of this year's installations, was utility-scale PV, rooftop was 4 GW, and 1.4 GW were off-grid capacity. Rajasthan and Gujarat are leading the country's installation rate.

By the end of 2023, India generated 14.4% of its electricity supply from renewable sources. The share keeps increasing fast but falls short of those of other major countries.

At year-end 2024, total solar capacity eventually reached some 100 GW.

It is interesting to note that India now equals Germany's total PV installation at the same time. But it stays currently well behind China, the other greatest country of the world with 1000 GW of PV now deployed in 2025.

By 2030, and that is soon, the country plans a total of 500 GW installed.

Solar continues to be the major contributor of renewable capacity in India with 44% of the total renewable mix, followed by the capacities of wind power and that of large hydro at 24% each.

India still imports more solar cells and modules than it exports. JMK reported that domestic PV manufacturers accounted for 66% of shipments and Chinese importers for the rest.

The 2024 Union Budget has a pivotal moment for India's solar industry and the government's commitment to renewable energy and sustainable development. The budget aims to accelerate solar adoption, enhance energy security, and create economic opportunities across the country.

One of the standout initiatives is the PM Surya Ghar Muft Bijli Yojana. It aims to install rooftop solar panels on 1 core (10 million) households. The scheme was decided by the government in February 2024. It provides a subsidy of 60% of the solar unit cost for systems up to 2 kW, or INR 30,000 subsidy for a 1 kW system, and INR 60,000 for a 2 kW system.

There is a strong interest in it, it is being said.

2.2.1.1.4.2 *Japan*

Japan is another solar energy pioneer.

Japan was the world's first country promoting grid connection of PV, the PV rooftop, and net metering on a large scale.

Initially, electric power companies in Japan were reluctant to open their grids for many small and distributed generators they could not directly manage. The issue was the stability of the grid. Therefore, in 1986, NEDO initiated a large demonstration project at Rokko Test Field at Kansai Electric Power Company KEPCO; it operated until 1990 a hundred house-like buildings with a 2 kW

PV system on top, all grid connected. It concluded after the test that the reliability and safety of the grid were maintained.

Then, the power industry accepted distributed PV systems to be connected to the grids. The PV industry, Kyocera, Sharp, Sanyo pushed things further. Kyocera build the first PV houses in the early 1990s.

The power companies started the net metering program in 1992.

After the German Ministry of Research in Bonn had launched in 1989/90 a 1000 PV roof subsidy program that served as a reference in Japan, MITI came up by end 1993 with a new PV subsidy program for PV rooftops. PV manufacturers rapidly got organized to start mass production for residential PV systems.

At the turn of the century, Japan was the world's number one solar PV player. Its market was the biggest and it was an important exporter of modules.

Since then things have changed. Nowadays, Japan imports most of the panels it uses. It is no longer the frontrunner of global PV deployment, but comes still fiercely in a group with India and Germany. In 2023, Japan installed 4 GW of new PV for a total of 87 GW. RTS, the rooftop sector, came for some 40%. 1 GW was residential PV. Together with the C&I sector, it is supported with the FIT.

In 2023, PV stood for 10% of the country's total power capacity. Together with hydro power and some wind power, the renewables secured 26% of all power capacity.

PV provided 4% of electricity generation well behind thermal power, hydro, and nuclear power.

In the perspective of 2030, Japan has decided the "Green Transformation GX." It targets, inter alia, 36–38% of renewables and 20–22% of nuclear power by then. **104–118 GW of PV,** 10 GW of offshore wind power are planned.

2.2.1.1.4.3 *South East Asia*

2.2.1.1.4.3.1 *Sinovoltaics SEA*

Considerable volumes of PV are produced in the southeastern countries of the continent.

In 2024, there were PV production factories in Cambodia, Indonesia, Laos, Malaysia, Singapore, Thailand, and Vietnam.

Hong Kong-based compliance and quality insurance company Sinovoltaics keeps track of them.

In 2024, a total module production capacity of 79 GW was identified there. From the supply chain, 50 GW are added for solar cells and 23 GW for silicon ingot production capacity.

2.2.1.1.4.3.2 *South Korea*

PV in South Korea is no longer what it used to be. In 2024, the country added some 2.7 GW of new PV capacity, bringing the total to some **30 GW**.

In 2023, PV had accounted for 18% of the country's overall power capacity and 6% of electricity generation.

But PV installation rates have been continuously declining since 2020 and the future outlook is negative as well. The reason is the country's energy policy with a drastic reduction of support for renewable energies on all levels. *Nuclear interests have now the favors of the government, leaving less room for support of solar power.*

2.2.1.1.4.3.3 *Vietnam and the ASEAN countries*

On international standards, the 10 ASEAN countries are unfortunately late in embarking on renewables' and PV deployment, preferring to cover their growing power needs with the conventional fossil sources. As an exception, only Vietnam has realized PV deployment at a relevant size.

For the year 2024, half of the electricity supply of Vietnam came from coal, a third from hydro, and a tenth from PV.

In 2024, Vietnam had some 19 GW of PV power deployed in total, PV provided 23% to the country's total capacity.

It had 100,000 PV rooftops for a total power of 9.5 GW.

In 2018, the country had virtually no PV or wind power installed. After that year, the installation jumped considerably upwards thanks to the introduction of a FIT for PV and net metering for rooftops.

After the FIT was discontinued in 2022, PV's deployment also came to a halt.

With its Power Development Plan VIII (PDP 8) from 2023, the country tries to shift away from the use of fossil energies. In 2024, two new decrees provided incentives for further solar deployment, such as the following:

- In July 2024, Decree 80 for "direct PPAs," DPPA as it is called, was issued. It comes for utility-scale PV generation.
- Then in October the same year, came Decree 135, furthering self-consumption via PV on rooftops. Tax incentives are available for support.

Figure 2.11 The author at the Gualong Bay, Vietnam, 2012.

Vietnam has with its PDP8 a great vision for its renewable energy supply by 2030:

- 73 GW of PV installed from 19 GW in 2024, providing 31% of the country's electricity supply, and
- 38 GW of wind power capacity for 16% of electricity supply.

Figure 2.12 The author in Hue, disguised as a Vietnamese Emperor.

2.2.1.1.5 PV in Africa

2.2.1.1.5.1 *MENA, Middle East and North Africa, "Vision 2030"*

The big solar and wind resources in the vast territory of MENA are so far by and large untapped. And there is no lack of finance to go ahead. Less than 20 GW of PV and even less of wind power capacity are installed so far, even though the installation rate has increased 40% over the last 10 years.

But now they are coming; tremendous projects of renewable energy installation are set going. Big installations for solar PV and wind power have been announced to cover the needs for supply of electricity, in particular for water desalination, irrigation, generation of green hydrogen, the latter to produce green ammonia and fertilizers thereof. Cable connections between the MENA countries, Europe and India are part of the effort to exchange and export renewable electricity.

Saudi Arabia

The country has announced "Vision 2030," aiming to have by 2030 50% of its power capacity, 58.7 GW that is, renewable: 40 GW of PV, 16 GW of wind power and a few others. By 2022, only 0.5 GW of PV and wind power were installed. A year later, 13 GW PV projects had already been launched, the biggest of 2 GW in the Mecca region. 1.8 GW of wind power was initiated, too. New PPAs trigger the renewable markets. A hydrogen development project, the world's largest, and blue ammonia play a prominent role.

NEOM, a $500 billion Mega city project for 9 million people is under development; it will be entirely powered by PV and wind with green hydrogen. It may be the flagship of Vision 2030, but experts estimate that it may need 100 years to build. By 2030, some 300 thousand people may be living there, not yet the millions announced, it is estimated.

United Arab Emirates

Dubai: Dubai in the United Arab Emirates is also building a new City: the WBR City in Dubai. Its cost is estimated at $80 billion.

Moreover, Dubai is developing a 5 GW solar park, to be ready by 2030. So far 0.9 GW are in operation, the next phase of 1.8 GW should be ready by 2026.

Abu Dhabi: It is the home of IRENA. With Dubai, the country is the most active in the region for Renewable energy development. MASDAR, the "future-energy company," is the country's arm for this purpose. In 2023, MASDAR invested some $30 billion in 40 countries or so.

MASDAR is highly interested in green hydrogen. 1 million ton production is planned for 2030. It plans also a $34 billion hydrogen project for Mauretania. The UAE wants, too, to become the world's first green aluminum producer.

A 2.5 GW cable connection with India is under consideration.

In January 2015, the president of the UAE launched with MASDAR in Abu Dhabi what may become the world's biggest PV plant providing uninterrupted supply of 1 GW every day in the year and every hour in the day. It will be a 5.2 GW PV plant associated with a 19 GWh BESS.

Well done!

Egypt

Since 2019, the country already has Benban, a 1.6 GW PV plant covering an area of 37 km^2.

Algeria

This vast country has less than 1 GW of PV installed, and even less in terms of wind power.

By 2030, it plans 3 GW PV rooftop, maybe Africa's largest, on the new airport of the city of Oran.

Tunisia

1.7 GW PV capacity is planned by the government, to be ready by 2027. Early 2025, to start with, four projects totaling 500 MW were approved after international tendering.

Morocco

In 2022, the country had 4 GW of PV installed and also some 2 GW of wind power. By 2030 for both together, some 10 GW are planned. 2 new PV plants covering 241 ha are designed to give the country by 2030 a 52% solar power capacity.

A new PV project was launched late 2024 in the Western Sahara desert with a high-voltage cable connection of 1400 km with Casablanca.

The Xlink, an undersea HVDC cable, 3800 km long, connecting the country with the UK to transfer solar- and wind-powered electricity is to be completed by 2030.

2.2.1.1.5.2 *Black Africa, "Mission 300"*

Still today, over 80% of the world's population without electricity live in "sub-Saharan Africa" as it is called. Their number is some 570 million (not thousand, millions!); over the past years, it was expanding instead of shrinking. In Congo, the vast RDC in the middle of the continent, even 80% people live without electricity. The problem is not at all new. We, the PV pioneers, have worked since solar PV was available to do something about it. But it's difficult.

Figure 2.13 The author lecturing PV at the German Goethe-Institut in Kano, Nigeria, in 2011.

Africa, a continent of conflicts and local wars everywhere.

Here has been announced the latest effort: getting electricity to 300 million people in 5 years: "*Mission 300*."

On January 28, 2025, meeting in Dar Es Salaam, 30 African heads of state and governments committed to expanding access to reliable, affordable and sustainable electricity to power economic growth, improve quality of life, and drive job creation across the continent.

Mission 300 unites governments, development banks, partners, philanthropies, and the private sector to connect 300 million Africans to electricity by 2030—providing renewable mini-grids and solar home systems. Partners pledged $50 billion in support of increasing energy access across Africa, it was said in a joint Declaration. That one will now be submitted to the African Union Summit.

The President of the World Bank Group, Ajay Banga, reminded "**Access to electricity is a fundamental human right.**" The World Bank Group WBG and the African Development Bank will allocate most of the funding of Mission 300. Also associated are the Rockefeller Foundation, the Energy Management Assistance

Program (ESMAP), the Global Energy Alliance for People and Planet (GEAPP), the Sustainable Energy for All (SEforALL), the Sustainable Fund for Africa, Distributed Access with Renewable Energy Scale Up (DARES) platform, IFC, MIGA and development partners from the WBG, Accelerating Sustainable and Clean Energy Access Transformation ASCENT, the Regional Emergency Solar Power Intervention Project RESPITE, West African Power Pool WAPP.

The whole Establishment for energy in Africa will participate. Let's hope this time it works. See You in 2030!

Ethiopia, a new hub for Renewable Energy in Africa

With over 100 million people, Ethiopia is a major African country. In March 2025, the fifth "Solar Africa" 2025 took place in Addis Abeba.

The Japanese Toyo Solar is producing solar cells in this country. By October 2025, an automated production line of 4 GW will be running at Hawassa. Some of the first produced in Africa! Besides solar cells, the whole value chain from silicon to modules is part of the production.

In February 2025, the government issued a call for tenders for two PV projects of 100 and 125 MW capacity.

And there is **hydro power**. In July 2025 a Grand Dam on the Blue Nile River, 6 GW of power, was completed. It took years to be built and will take more years in the future to fill it up. As one could expect, it is much disputed by all neighboring states, in particular Egypt.

2.2.1.1.5.3 *South Africa*

The country is today by far the African continent's most important solar market, at least in the sub-Saharan part of it.

In 2024, some 8.3 GW in total stood deployed there, 2.5 GW utility-scale and 5.8 GW rooftop RTS. It is 5% of the energy mix.

The country employs primarily coal for its electricity generation. But just recently, there has been a real solar boom triggered by house owners, who want to escape the daily power cuts. The latest trend goes for the combination of the PV plant on the roof with a

BESS, a lithium-battery BTM, behind the meter, on the user's side of the connection.

The country adopted in 2025 SAREM, the SA Renewable Energy Master Plan. Up to 5 GW of new capacity is to be grid connected annually in the next few years.

2.2.1.1.6 *PV in Australia*

The fifth continent is a great solar country. In 2024, it installed some 6 GW, thus reaching a total of 40 GW of PV. That is a lot less than Japan's PV capacity, as we saw here above. Australia is blessed with a great solar resource but in terms of population density it is a very small country. At over 1 kW of solar PV installed per capita, Australia holds together with Holland in Europe a world record. In every other respect, both are the contrary of each other: Holland is a comparatively very small country with a high population density and a modest solar resource while Australia is from all these criteria to a large extent the opposite.

The solar boom in Australia is very recent; it all started just 20 years ago. Nowadays, the country counts well over 4 million PV plants.

Over 50% of total PV capacity is installed rooftop, called RTS. 30% of all households benefit from PV rooftops. Distributed PV is supported via a FIT.

Just in 2023 57,000 batteries BESS have been installed with PV plants. Those are installed behind the Meter.

The renewables are doing well in terms of electricity generation; in 2023, they secured 35% of it; PV 16% with 11% from RTS, wind power 12%, the rest hydro.

The growth of PV deployment is accelerating. The consultancy Ember found that in October 2024 PV provided 25% of electricity supply while that from coal passed below 50%.

By 2030, the country plans 120 GW of PV+ wind power capacity.

Eventually, one has to mention the project **Sun Cable.** The project of $13.5 billion will be financed, it is said, from private sources. It is an enormous solar plant with storage to be installed in the northwestern territory. Electricity from 4 GW of the PV power may go to some 3 million homes in the country and from

2 other GW the current may go by an undersea cable to Singapore. The cable is supposed to be 4300 km long.

Australia's government has given its approval at year-end 2024. Other approvals are expected until 2027 and start of production by 2030.

1987: The start of the world's first rooftop PV in Australia. On cars as it were, not on buildings.

Australia is a big country, a whole continent; but in terms of population, it is just a fifth of Japan, a country that occupies only a few islands. Australia is not only very big in size, but it also benefits from an extraordinary solar resource.

Some early solar fanatics thought to take advantage of the country's privileged situation to prove the viability of road traffic only powered by the Sun's radiation, no doubt a new challenge. A major demonstration of it was the first transcontinental World Solar Challenge (WSC) with a race through 3005 km across Australia, all the way from North to South. It was a route already explored in earlier years by some pioneers.

It is important to note that this challenge in 1987 to demonstrate PV rooftops on cars came earlier than Japan's first PV applications encouraged on grid-connected building, which started PV's mass markets in modern times.

A lot of the established car makers took part in the WSC, GM with a team of 91 persons from 13 different auto firms, Ford, later Honda in another WSC race with 11 teams, and not to forget some enthusiastic university teams from Switzerland, Germany, Pakistan with the Solar Samba, the United States, Australia, Canada, the UK.

All cars were different from each other in design. Their common feature was PV on the roof and a storage battery, no other engine.

What a big show all around early solar PV! "Creating a SOLUTION not POLLUTION."

As Gallus Cadonau from Zurich reported in an article on solar races, the first WSC started on November 6, 1987, in Darwin. On day 5, after 44 hours and 54 minutes, the Sunracer of GM crossed the finish line in Adelaide after an average speed 66.9 km/h.

2.2.1.1.7 *PV markets in Latin America*

OLADE, the Energy Organization of the 27 countries of Latin America and the Caribbean, reported that 79% of all new power installations in 2024 were of the renewable type. A majority of power in Latin America is renewable today.

Hydro is the leading renewable power with 37% of the total installed, followed by wind power with 66 GW and 12% of the total. PV capacity increased from 60 GW in 2023 to 80 GW in 2024, 14% of the total.

PV-generated electricity met 7.4% of demand in 2024, up from 6% in 2023. Wind power in 2024 generated 12% of all electricity supply.

2.2.1.1.7.1 *Brazil*

Brazil was one of the countries suffering the most from atmospheric pollution in 2024, a devastating year for South America. Following CAMS, the Copernicus Atmospheric Monitoring System, the country suffered from a severe drought, resulting in a lot of wildfires. 18 million ha of land have burnt, a lot more than in the previous year. Clouds of dust covered parts of the country, including big cities like Sao Paulo, for long periods with pollution. Mega tons of carbon were released.

After the bad news comes the good news: 50 GW of PV installation has been reached in October 2024, + 13 GW in the year so far. For year-end 2024, some 52 GW of PV were reached, almost 20 GW of new solar capacity in the year. 37.4 GW of it were distributed ones and the rest 17.6 GW large scale. At last, in the first trimester 2025 55 GW of PV capacity in total got installed.

Sao Paulo ranks as the leader of solar rooftop capacity.

Solar panels are imported and remain cheap despite an increase up to 25% of import tariffs in November 2024. Prices dropped by some 50% in 2023. Financial payback time is 4 to 5 years. For financing, credit cooperatives are also available. New net metering is organized.

For its energy consumption, Brazil is already a "clean" country today as it relies for over 90% of its power capacity on the renewables, 25% for PV. In terms of electricity generation, PV comes third with almost 8% of the total supply, after hydro with 60% and wind power.

Solar power contributes to Brazil's economic development. It has created 1.4 million jobs.

New government programs are set up for the underprivileged. In the Amazon rainforest, mini-grids for 3 million people are in operation using diesel power. Their conversion to PV is hindered by the need of open spaces instead of woodland for PV.

2.2.1.1.7.2 *Peru*

In 2024 was started at Joya de Arequipa the Solar Plant Fotovoltaica Illa by the Spanish Enhol in a PPA with the utility Kallpa. **It has a capacity of 472 MW_{dc}/396 MW_{ac} to be completed by 2026.** The plant uses 6800 Solar trackers provided by Solar Steel and is supposed to generate 1.2 TWh annually when completed.

2.2.1.1.7.3 *Uruguay*

This small country is well known for its Renewable Energy Revolution. By now already, virtually all its electricity supply comes from renewable power. Wind power installed is as high as that from hydro power, both come 1.5 GW. Bio-energy stands for 0.7 GW and PV has emerged as **330 MW.**

2.2.1.1.7.4 *Chile, Atacama*

In 2024, Chile had 13.8 GW of PV installed in total, PV provided over 20% of electricity supply in 2024. Like in many other countries, PV is the champion of capacity growth.

A particularity of PV in Chile is the low part of rooftop with 3% of the total. There is, however, a great interest for PV from farmers; it is for electricity supply and their water wells.

Already in 2022, PV and wind power overtook coal in terms of electricity supply. The share of coal decreased dramatically since 2018 in Chile.

As a result of the steep increase in renewable electricity consumption, the grid cannot keep up and becomes overloaded. Solving the problems of curtailment via BESS is high on the political agenda.

Installing PV in the Atacama desert attracted early attention of investors. Nowadays, it may rain a bit in the Sahara, not so in the Atacama desert, the driest place on Earth. It is known for its observatories, taking advantage of the clear sky there.

European utilities such as Enel and EDF started to invest in several very big PV plants. One of the latest was the 480 MW Atacama Desert Solar Park near Antofagasta commissioned in 2024.

The French Engie started operation in 2024 of its 124 MW plant also near Antofagasta; it was installed at the site of a decommissioned diesel plant. It is associated with a BESS of 48 MW/418 MWh, a storage coming in 152 containers.

And even better, the US Contour Global has started there a 221 MW PV plant associated with batteries of 1200 MWh. And there are others...

The Spanish Grenergy is building the project Oasis de Atacama with *2 GW of PV and 11 GWh of batteries from the Chinese CATL.*

2.2.1.2 Latest trends in PV technology

2.2.1.2.1 *The world of silicon*

The 2020s witnessed a radical change in silicon solar cell technology, primarily from PERC to TOPCon.

By 2022, PERC at last conquered the world's silicon cell markets with a share of 89% of production.

But by 2024, it had lost it already again in favor of TOPCon, which raised its share to 81%, leaving PERC with 7% far behind. And those technologies are much different from one another.

As mentioned earlier, modern TOPCon solar cells and their modules have conversion efficiency, on large industrial scale, of some 26%.

Dong Xiaoqing, director of New Energy Design Institute, highlighted five advantages of TOPCon: high efficiency, low degradation rate, outstanding temperature coefficient, excellent low-light performance, and strong site adaptability. She mentioned this at a speech at China's PVIA, the Industry Association, on the

"Development Trends of TOPCon Solar Cell Technology" in August 2024.

TOPCon is outstanding for its high bifaciality and low-irradiance performance, it was concluded.

The PERC technology is rather an extension of classical silicon cell technology; not so TOPCon, which introduces nano-technology into the PV sector.

Here is an overview of the PERC cell:

PERC solar cells are an improved version of conventional solar cells. They have a better conversion efficiency of up to 10% over the traditional ones.

PERC stands for Passivated Emitter Rear Cell. The ***p-type*** *crystalline silicon cell (boron doped) has a passivation layer at the back to minimize recombination losses and increase light absorption. Thus light not yet absorbed is reflected back inside the cell. It gets a second chance to be absorbed and augment the cell's efficiency.*

The PERC technology has been developed in 1983 by M. Green and co-workers at UNSW in Australia.

Here is the structure of a TOPCon solar cell:

TOPCon stands for Tunnel Oxide Passivated Contact. It puts the world's PV technology upside down as it employs phosphorous-doped ***n-type*** *silicon and not p-type as PERC and the classical cell structures do.*

It was developed much later than PERC, at ISE of FhG in Freiburg, Germany. The technology was first published in a paper in 2013 authored by F. Feldmann, S. Glunz, and M. Hermle.

The principle of TOPCon is that the metal of the connection contacts does not touch the silicon, to avoid recombination of the charge carriers. On the top comes a highly doped p-layer with the n-type crystal, forming the p-n transition that creates the PV effect.

On the back a very thin layer of silicon oxide, just a few nanometers, or atomic layers thick, to allow the current to pass by the tunnel effect, quantum mechanically, is in contact with the crystal. Under this "tunnel layer," another layer of p-doped silicon is added to contact the electrode.

n-Type cells promise higher efficiencies than the traditional p-type ones. With p-type cells, there are inter alia losses when sunlight hits the module for the first time, the initial degradation. Other advantages of n-type cells include a higher temperature resistance and a lower loss of efficiency with increasing temperature. But n-type cells are more difficult to manufacture.

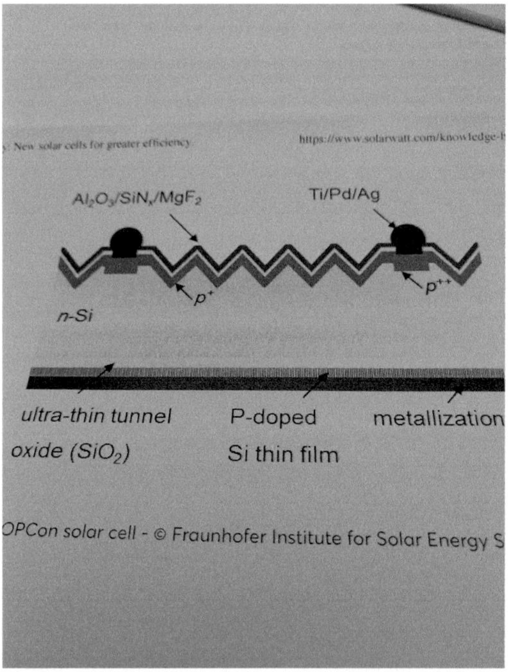

Figure 2.14 The TOPCon silicon solar cell, as invented by the FhG ISE in Freiburg, Germany.

When ISE was started by the late Adolf Götzberger in 1981, the author was a member of its first Board.

Later on, the Chinese industry developed the technology further by introducing bifaciality. As usual, one creates on the front a boron-diffused emitter on the n-type crystal. On the back comes the tunnel layer of SiO_x /p+poly. Both the front and the back can create a PV effect and generate electricity.

The German association VDMA, in its **ITRPV report 2024, the International Technology Roadmap for PV, confirms that**

- By now all silicon cells are monocrystalline; polycrystalline silicon has completely disappeared from the market.
- TOPCon is the dominating technology.
- Bifacial solar cells keep a market share of 90%.
- "M10" wafers of 182 mm^2 dominate the market.
- At 40-year lifetime, modules stay above the usual warranty period.
- For rooftop, modules of an area of 1.8 m^2 up to 2 m^2 are currently the preferred ones.
- Modules up to 3 m^2 dominate the utility-scale segment.
- Most new manufacturing plants have an annual production capacity of 5 GW and above, but smaller ones even below 1 GW capacity are planned for niche markets.

In 2025 and for some years to come, a few other PV technologies are established on the markets next to TOPCon and some PERC; their share on the international markets currently stays below 12%:

- Various types of back contact cells xBC.
- Hetero-junction cells HJT employing crystalline silicon with an amorphous silicon layer on the front and the back, respectively.
- CdTe thin-film cells.

Already in 2024 came the Chinese manufacturer Longi with another new type of cell, combining some of the above features, a *silicon hetero-junction BC cell, back contact by laser sputtering.* They call it MI-MO 9 and claim an efficiency of 27.3%, better than the TOPCon.

Details were given by a paper published by Wu H., Ye F. Yang M., et al. in *Nature* 635 (604–609), October 1, 2024.

In April 2025, Longi reported to have reached an agreement in Greece for the deployment of 226 MW with its HI-MO 9 BC modules, of 660 W each and 24.4% efficiency.

2.2.1.2.2 *Global leaders of the solar cell and module industry*

Following my friends from RTS Corp. in Tokyo, the following 10 firms are the world leaders of current PV markets:

- **JinkoSolar (it was at year-end 2024 the first to announce having 300 GW of PV modules produced and sold since 2011, a world record; its Tiger Neo n-type is said to be the best-selling module)**
- **Longi Green Energy**
- **Trina Solar**
- **JA Solar Technology**
- **TW Tongwei Group**
- **Canadian Solar**
- **Chint New Energy**
- **Risen Energy**
- **DAS Solar**
- **GCL SI**

All the firms on this list are Chinese. **They all produce TOPCon cells and modules.**

HJT modules are offered in China by Risen Energy in parallel to TOPCon, and by Tongwei, too. And as BC cells by Longi, see above.

In India, REC India offers HJT modules; so does Meyer Burger in Europe and the US–the latter declared bankruptcy in 2025.

SNEC 2025, World's largest PV Exhibition and Conference in Shanghai, China: This yearly exhibition and conference was the 18th. It took place like all the previous ones in the month of June. I had attended one of the first ones, with drums and trumpets, but not this one.

SNEC 2025 was attended by half a million people from 95 countries! My friends from RTS Corp in Tokyo were among the visitors and wrote a nice report about it.

They wrote: In solar cell technology, p-type crystalline silicon PERC cells have almost disappeared and n-type TOPCon cells have become mainstream. From the perspective of cost and reliability, they are expected to remain at the center of the market for the foreseeable future.

An increasing number of TOPCon solar cells are now large, 210 mm in size. The leading manufacturers are all those mentioned already above in this chapter. Module sizes range from 660 to 800 W, highest efficiency for Longi at 27.81%. Most modules on offer employ 132 cells of 210 mm cut in half.

2.2.1.2.3 *Outsiders*

2.2.1.2.3.1 *Thin-film cells and modules*

In the early days, when PV was developed in earnest after 1960 or so, it was for us, the pioneers at that time, evidence that for massive deployment in the future the PV technology had to be thin-film, not the silicon crystalline.

Leaving alone the fact that silicon cells already existed and had a real market in the space business, while the thin-films existed only as research items. For reasons of fundamental physics silicon solar cells had always to be some 100 micrometers thick, coming with all the headaches of purifying the silicon ingots at high temperatures first, drawing crystals, cut them into wafers, the doping, etc.

Thin-film cells can be a 100 times thinner and modules can be obtained by simple spray on a support. And in theory, thin-film cells like those of CdTe have a higher efficiency than silicon cells.

Hence, the excitement about the new silicon cells and their promises had not yet cooled down that new materials for thin-film cells indeed attracted the attention of US scientists, GaAs, GaTe, and above all CdS: Rappaport published proudly that they had reached 4.5% efficiency with it.

But eventually it did not work that way. Today's PV market is occupied by silicon and First Solar in the US keeps only a small part with its CdTe technology.

- **CdS/Cu_2S was the preferred early contender.** A number of groups around the world worked on it, in the United States,

Germany and France among others. I had passed a thesis on the material in 1965 and later led the French development effort on it in half a dozen groups. One after the other, global efforts were given up as stability problems looked insurmountable

Photon Power

I was contacted in 1974 by **John F. Jordan,** a Vice President of D.H. Baldwin, the builder of pianos and musical instruments in Cincinnati, Ohio. Working on optical sensors for such instruments they had discovered CdS and Photovoltaics. The company had offices in El Paso, on the Rio Grande in Texas. On the opposite side of the river is Juàres, Mexico: to the North, the city is surrounded by hundreds of miles of deserts.

John had developed the idea to produce all the power needs of the United States by covering these deserts by CdS Solar panels. Hundreds or thousands of GW; he was ahead of his time, as large-scale PV came only some 50 years later, didn't it.

To this end, he wanted to use an innovative process to produce the thin-film cells on a large scale. Pilkington in the UK had just developed a new process to produce flat glass: pulling the liquid glass over a liquid tin-bath at 500°C. John's idea was to spray the CdS from a liquid on the hot glass and a thin conductive layer. The technology did already exist at the time and was patented.

I was in those years widely known for my interest in CdS Solar modules, and I got enthusiastic about John's ideas when we met. We became good friends.

From my side I was in contact with the French authorities, and with the Oil Company Total; in particular my friend François Fiatte, in charge of new activities there. Eventually was formed in 1975 with my help the company *Photon Power* in El Paso, owned by Total and Baldwin and managed by a director from London, UK.

$20 Million were newly invested. The process worked and up to 8% efficiency was achieved on some modules produced; were there not the few pinholes that short-circuited the very thin cells that are only a micron thick. The project failed as it turned out to be too innovative and also owing to a lack of personnel to carry it through. The director at the plant had family problems, John got close to 80 years old, and I had a new job at the EU in Brussels.

Figure 2.15 My good friend Dieter Bonnet (+), the German developer of CdTe solar modules.

- **A similar material to CdS/Cu_2S is CdTe.** It was the follow-up material to attract global interest. By today it is the only one to have survived and passed into industrial production in the United States.
- **Then came amorphous silicon, developed first in Europe and the United States.** Later Japanese firms industrialized it: the first pocket calculators employed it. **It has survived in association with the HJT cells.**
- **Followed CIS and CIGS solar cells.** They entered industrial production by Nanosolar, Solyndra and Solar Frontier in Japan. Everything was eventually aborted.
- **Nowadays comes ultrathin perovskite.** It is the favorite subject of a new generation of researchers. Dozens of groups around the world are trying to develop it.

2.2.1.2.3.2 *R&D attempts*

Perovskite is the material that attracts most R&D efforts on PV around the world today. It is an interesting material, comes in many types of compositions, organic or inorganic; is studied, too, for its optoelectronic properties, as light emitting diodes.

Solar cells of it were realized first in 2009.

Perovskite consists basically of methylammonium metal halides in many varied compounds. On very small areas by now conversion efficiencies of some 26% have been reached.

In tandem cells in combination with silicon even 34% efficiencies have been reported. Perovskite layers can be made very thin, even below a micrometer thick. They absorb solar light more in the blue part of the spectrum compared to silicon; this makes the option of both combined in tandem cells attractive.

By today it is not yet industrialized in any way. Layers are very easy to realize. But currently only cells on very small areas have been produced successfully. The problem as usual for thin-film cells is short- and long-term stability.

There is much hope for a breakthrough soon of tandem perovskite + silicon cells. Future will show...

2.2.1.3 PV building integration, solar architecture

Figure 2.16 PV helps integrating buildings in their natural environment.

PV is part of the purpose of making buildings attractive, pleasant, and comfortable for living and working in them. In many countries, PV rooftop increases the commercial value of a building. New research in 2025, reported by "EnergySage," revealed that PV can boost the sale value of a house by thousands of dollars. On the 5000 houses in California that have been analyzed costing $790,000 on average, solar PV did increase their property value by 5% to 10%, more than double previous estimates.

PV is an essential component for an attractive EPB, an "Energy Performance of the Building," that displaces pollution and CO_2 emissions from oil and gas heating with those derived from the Sun.

Solar architecture is the purest form of achieving heating and cooling in the building design and structure by Solar radiation, associated as it were, with PV rooftop and façade integrated. (See separate Chapter by Stefan Behling)

Heat pumps

Heat pumps for heating and cooling combine perfectly with PV to make a building, whatever its purpose, an important part, if not totally, for its electricity demand + its heating or cooling needs independent from other sources than local solar radiation.

PV + heat pump is the modern way meeting a building's energy needs.

Since 2010 or so, the heat pump market has developed in parallel with the global PV market and like PV with an extraordinary speed. As of 2024 some 190 million heat pumps are deployed globally in many countries around the globe. The potential is huge, by 2030 a world market of 600 million units is projected. Eventually 80% of all heating and cooling demand in the world's buildings could be met with heat pumps.

Year-end 2024, some 26 million heat pumps were operating in Europe, 2 million more than a year earlier.

But like for PV, thus far China has reaped the market crown for heat pump installations. China comes currently first with a share of over 25% of global installation, followed by North America and Europe. In China, PV + heat pump are already by now in many regions the norm in new buildings.

In China, a heat pump with PV may have a 6-year payback period.

"Passive houses," or better "Passive Solar homes," were quite popular some years ago, around the turn of the century.

Interest has by and large vanished since then with the arrival of the heat pump market and PV on residences. But the basic features of bioclimatic design of buildings kept more than ever its importance in modern architecture, too.

The aims of keeping the house air and watertight, use heat exchangers for fresh air intake that are essential in the passive house design have had their day.

But important features of general value, in particular for energy efficiency, and rooftop PV, are part of modern EPB standards, such as:

- **Transparent insulation: today double or triple glass windows,** once a research item for plastic honeycomb structures or other insulation products for windows and walls
- **Insulation against heat losses through the soil, the walls and through the roof**
- **Thermal mass to store heat such as high-performance walls**
- **Window shading to avoid overheating**
- **Condensing boiler for auxiliary heating**

Maybe a Solar water heater, Sunspaces...

For those who want to be a 100% supplied by solar electricity all year round, one may mention HPS, the company Home Power Solutions offering picea. That way the electricity generated by PV (a lot of it for this purpose) on the building can be stored by electrolysis-generated hydrogen. And so, interseasonal storage is provided offering solar power any time.

Such a system that includes also a BESS for short-term storage is offered today for €100,000 and above...

2.2.1.4 A fascinating history of PV development since Becquerel in 1839

The PV effect was first observed in 1839 in liquids—semiconductors didn't yet exist at the time. The discoverer was young Edmond Becquerel when playing around with a Voltaic cell. (Alessandro Volta had invented the Voltaic pile that he showed to Napoleon in 1801—the Emperor was impressed and made him a count.)

Becquerel's discovery was found important enough to be published in a scientific journal, *Les Comptes Rendus de l'Académie des Sciences* in Paris. It reported "the production of an electric current when two plates of platinum or gold immersed in an acid, neutral, or alkaline solution are exposed in an uneven way to solar radiation."

The building in Paris where the discovery was made still exists. It even can be found in a nice environment at the "Jardin des Plantes" not far from the city center of Paris. Just recently a commemorative plaque of the event that happened there and was about to change the world some 200 years later was attached near the entrance of the building. The discovery had long been forgotten by the Parisians who had preferred to erect at the building a statue of his son, the discoverer of radioactivity who had worked in the same building.

But now Edmond found its place there under the care of D. Lincot. It happened that Lincot was also the 2024 winner of the E. Becquerel Prize. It is the most prestigious Prize for merits on photovoltaics in Europe that I created myself on behalf of the EU in 1989, 150 years after the discovery. (The Prize is presented to each year's winner at the festive openings of the EU PVSEC, another initiative from 1977 by me on behalf of the EU.)

Among all the outstanding scientists and personalities who once received the Becquerel Award for merits in PV, I want to mention the following:

- Roger van Overstraeten, the founder of IMEC in Belgium
- Werner Bloss from Stuttgart
- Antonio Luque from Madrid

- Mort Prince from Washington, co-worker of the discoverers of PV in silicon at Bell Labs
- Karl Heinz Krebs from the EU labs at ISPRA, Italy
- Adolf Götzberger from Freiburg
- Walter Sandtner from Bonn, the pioneer of PV integration in buildings
- Viacheslav Andreev, St. Petersburg, Russia
- Masafumi Yamaguchi, Toyota
- Richard Swanson, founder of SunPower in the United States
- Mechtild Rothe, EU Parliament

In 1849, 10 years after Becquerel's discovery, a certain Alfred Smee in London coined the term "photovoltaic," PV.

In 1886, Charles Fritts, the inventor of the first solid-state solar cell was quoted as saying "Solar Energy will pour down on us long after we have run out of fossil fuels."

In 1891, R. Appleyard expressed his vision in the *Telegraphic Journal*: "The blessed vision of the Sun... by means of photoelectric cells... these powers gathered into electrical storehouses... to the total extinction of the steam engine and the utter repression of smoke."

As mentioned before, the American Charles Fritts invented the first PV cell 1883 on a semiconductor, selenium. Those cells had very low efficiency of 0.6% but were produced already on a commercial scale. Selenium cells had a market for decades despite their low efficiency until 1956 or so when the silicon solar cells came eventually to the markets. They had been used for camera exposure meters, photo switches, photoelectric eyes.

Until the 1960s, cameras in Germany came with photoelectric sensors of selenium—using an external battery, not PV—for measuring light intensity.

Silicon solar cells came much later; first had to be solved the question of gaining a material of sufficient purity, not an easy task. But still, Bell Labs in the US went ahead on the way of producing efficient silicon solar cells. **In** 1941, Russell Ohl at Bell filed a patent for a silicon cell and tried to produce one despite this lack of material that was pure enough.

Calvin Fuller successfully realized silicon crystals of sufficient purity in 1953. Doping one with boron led to the first practical silicon solar cell at Bell that year. It was published in 1954: "D. Chapin, C. Fuller, G. Pearson, A new silicon p-n junction photocell for converting solar radiation into electrical power. *J. Appl. Physics* 25, 676–677, 1054."

The *New York Times* had this event in 1954 on its front page. The first commercial use came in 1955 in telephones, for power transducers—as they said. For a panel 432 cells were connected to get 10 Watts at 22 V, for connection to a Ni-Cd battery. It was clear that this was not competitive, at best for 1 W applications.

Fortunately, in 1958 the space satellite market opened with the Vanguard satellite. And the Van Allen belts and their effect on silicon solar cells was discovered.

On the terrestrial markets, PV cost came $500/W in 1959—today it is 2000 times less! The silicon solar cell conversion efficiency for sunlight stood then at 8.5%. Shortly after in 1960, there came technological progress when grid fingers were applied on the cell's surface to reduce current resistance: efficiencies went up to 14% and costs came down to $175/W.

Since 1960, off-grid markets became of interest, first of them wireless communication. Other terrestrial markets of interest were light and loudspeakers in small villages, emergency systems on highways, safety satellite systems for ambulances, etc.

For all the reasons discussed previously, in particular the need to preserve environment and climate, what was needed obviously in the longer term was a mass market of PV beating all other power providers in terms of cost and competitiveness.

2.2.1.4.1 *The 1970s, the American pioneers, Analyzing PV's potential*

Right from the beginning in this attempt came the United States: In June 1971, the US President issued a message calling for programs to ensure adequate supplies of clean energy for the years ahead. Under the authority of the President's Office of Science and Technology, a solar energy panel was jointly organized in 1972 by NASA and the National Science Foundation NSF. Martin Wolf chaired the sub-panel on PV. The declared goal was a million-fold expansion of PV production rates and the attendant automation.

Contrary to the existing projections of the time, the Panel recommended to the US government a target of 20% Solar energy contribution to the US total needs by 2020, equivalent to the total energy consumed in the United states in 1970.

Later in 1973 the Californian Jet Propulsion Laboratory JPL held the famous PV workshop at *Cherry Hill*, New Jersey. It was sponsored by the NSF. A projection of large-scale deployment of PV was the subject of that important meeting.

The result of the Panel on silicon solar cells at that workshop, chaired by Paul Rappaport, later the founder of the NREL at Golden, Col., identified at last the potential of silicon cells for mass production that they got by today:

At a cost $0.50/Watt solar cells could be produced in future at a manufacturing volume of 0.5 GW annually; or even for $0.10/Watt at a production capacity of 50 GW. The probability of success were high as there is abundant material cheaply available, theory and technology are well understood and reliability is proven. Now some 50 years later that is exactly what the world has achieved in PV development—with *silicon* solar panels, as it were.

2.2.1.4.2 *France, Germany, the UK, and Brussels' EU Commission: Analyzing PV's potential in Europe*

In France, the "Groupe de Bellevue" published in 1978 "ALTER, study of an energy future for France based on Renewable energies." Participated experts from CNRS, Collège de France, INRA, the National Institute for Agricultural Research, and led by Ph. Chartier from ADEME, the Agency for Energy. The study concluded that in the perspective of 2050 it would be possible to change progressively in France from an energy system dominated by fossil energies to a stable autonomous energy system based exclusively on renewable sources, Solar energy in all its forms.

The study projected a maximum supply of electricity, in particular for residential use, of which 53% to be generated from PV. 80% of the fuel needs to be generated from hydrogen.

In Germany, a particular extensive work was performed and published by the Bundestag in Berlin, the German Parliament. Subject was the country's energy perspective by the year 2050. For the purpose a special Enquête Commission had been formed that worked regularly from 1999 to 2002. I was myself a member of it.

After a strong opposition from the conservative parties, known to be in support of atomic power, the official conclusion of this multiparty investigation, "enquête" in new German, was "A sustainable energy supply based on energy saving, efficiency, and renewable energies can be realized in Germany by 2050." The main author of the study report was my friend Harry Lehmann.

In the UK, a study by BP led by Tim Bruton on up-scaling of PV manufacturing concluded that for a silicon cell production line of 0.5 GW a module cost of €1/Watt was achievable without any technical breakthrough. In those days, BP was more than it is today a major player in the PV business.

The EU Commission in 1994 when I was myself leader of its Renewable energy Division projected in an "APAS" study for Europe in 2030 an installation of a PV capacity of 155 GW. A correct view, though it will, as we know by now, a multiple of it; but at the time of the study Europe had not yet installed its first GW.

2.2.1.4.3 *1973 Paris: The World Congress "The Sun in Service of Mankind"*

In the first week of July 1973 was organized by the French Associations AFEDES and COMPLES at the UNESCO House in Paris the International Solar Congress with over a thousand participants from all over the world, pioneers from the US, Europe, Russia. It was a powerful demonstration of everybody's wish to see the Sun's energy implemented without waiting any further.

All issues of possible solar systems were covered, in particular the heating applications, solar houses, or small solar engines for the need in villages in Africa. **A major issue was PV; a part organized and chaired by me;** I was in those days leading from Paris the French PV activities at the National Space Agency CNES.

2.2.1.4.4 *The 1990s, Germany, Japan: First steps into massive PV deployment*

Until those days, the applications of PV were totally restricted to stand-alone ones. They were not, however, although it could have been, for providing power and light to all those who don't have any; and there were over 3 billion people of those in need in the developing world. However, unfortunately in the meantime, by now a major part of the new power in the remote villages was installed with the polluting fossil fuel plants. The unique opportunity to supply billions of people with solar light, a potential mass market that would have initiated the low-cost potential of PV, has been missed.

What happened in reality was to supply PV to grid-connected buildings, and integrate them, too, into those grids. The industrialized countries took up the challenge to reduce their dependence from coal and all fossil fuels.

PV on grid-connected houses started exactly in 1989 and developed at a grand speed in parallel with PV's own development—and that of Wind Power and the other renewable power providers.

In *Germany* was indeed started in 1989 a program to install PV with buildings that are grid connected. It may sound incredible today, but it was one of the world's first!

Walter Sandtner started in 1989 on behalf of the Ministry BMFT for Research and Technology the "One Thousand PV Roofs Programme." Sandtner, newly appointed head of the Renewable energy division at the Ministry, felt at that time that the German public and the politicians requested with growing impatience that after 15 years of intense research on solar energy, something should eventually become visible.

In the end, some 2200 PV plants were deployed. They were spread over the whole country in cooperation of all regions and some research institutes. Before, there existed in total only six buildings in Germany with PV attached. Interestingly, the consulted experts had told Sandtner, in a three-page report, that such a program could not work. He replied then that the decision was already taken and they were only asked to give advice on how to implement it.

Everybody in Germany could make an offer after a call had been widely publicized, but there were regional quotas. The PV plants were 70% subsidized, part of it by the Ministry, a minor part by the Länder of installation. A condition was that the house was grid connected and the plant became rooftop. Aesthetic considerations were of great importance. Prizes were given to the most beautiful solar roofs.

Already after a couple of months, there were 40 different inverter types on the market, DIN norms were formulated. There were many new scientific studies and improvements. The price of a PV installation fell by a third during the program period. PV panels had not to be only German; American ones, Japanese, British, and Spanish modules have actually been used. On the former Munich airport, now the city's Trade Fair Centre, a 1 MW plant was installed. In cooperation with Spain, near Toledo, another 1 MW PV plant was erected. Other cooperations came up with more European countries, with the US, Russia, Brazil, Argentina, India, China, and Indonesia.

Importantly, the thousand-roof program proved that PV works also in more northern latitudes. And there was no need for additional land occupation.

The program proved eventually a big success. And it opened at last the door to massive PV deployment.

Japan was the country to follow immediately the good news from Germany giving the evidence that utilities can handle a widely distributed number of small and intermittent power suppliers on their grids.

The Ministry MITI came up with a new subsidy program for PV rooftops since 1993.

Before, in 1992, the utilities in Japan started net metering in support of PV, another revolutionary support of grid-connected building installations of PV.

Japanese PV module producers, Kyocera, Sharp, Sanyo got ready for mass production. By the turn of the century half of the world's panel production was Japanese.

In Germany, the political situation changed for the renewables in 1998 when the SPD, the social democratic party, won the federal election. Hermann Scheer, the influential speaker for his party on the Renewables in the Bundestag, the Parliament, saw then that the time was ripe for massive deployment of PV in Germany. To this end, the country needed a 100,000 PV roofs program, a smaller one could not really help. That's also what he told Mr. Sandtner.

After H. Scheer's party had won the government, he could come ahead with his 100,000-roofs program, as an extension of the 1000-roofs program of 1989/90. After having convinced Oskar Lafontaine, the new Minister of Finance, to go ahead with it, as well as the federal bank KfW, the program was started January 1, 1999. It was a program of subsidized loans for PV provided by the KfW.

It provided in support of the PV installation a reduction of 4.5% in the interest rate for a possible loan of up to some €6.75/kW for systems below 5 kW capacity.

The program was less successful than planned and ended in 2003 without a further extension. Altogether, 55,000 PV roofs were supported in the program for a total of 261 MW.

2.2.1.4.5 *Germany introducing the FIT (feed-in-tariff) for PV in 2000/2001*

The PURPA by President Carter of 1978 already had a provision obliging utilities to buy electricity from independent power producers at avoided cost—where avoided cost was left to interpretation...

Leaving alone the US PURPA, in Europe, the FIT started in 1988 in Denmark. At that time, the Danish government skipped the provision of a subsidy of 30% to the capital cost of Wind Power. In turn was introduced the FIT to the benefit of wind-powered electricity. It obliged the utilities to connect Wind turbines to their grid and buy the electricity at a fair price.

In 1990, the German Bundestag adopted the FIT for the renewable electricity (in practice it was only for Wind and small hydro-generated electricity) in Germany. To this end, H. Scheer had to organize cooperation also with Members of the Christian Democrats who had by then the majority in Parliament. The Feed-in Law, (called Stromeinspeisungsgesetz) as adopted then, guaranteed independent electricity producers a feed-in price of between 75 and 90% of the underlying electricity price.

Consequently, wind power development took off in Germany in 1990. But for PV, the FIT was not attractive enough.

And how did the FIT for PV start in Germany?

Actually for PV the FIT started in Germany's imperial city of Aachen. Wolf von Fabeck founded in Aachen the SFV, the "Solarenergie Förderverein" to promote the "cost covering price" of PV. SFV succeeded after a tremendous effort turning public opinion in favor of it. It got approval in 1995 by the City Council as the "Aachener Modell." It became effective when STAWAG, the Stadtwerke (local utility) followed and applied it in practice. In the meantime, Hammelburg (the town where H. J. Fell pushed it through) and Freising in Bavaria had already adopted the same model and many other towns followed.

PV electricity bought by the municipal utilities was paid 1.80 DM per kWh (almost €1), as it was the cost-covering price at the time.

The national FIT for PV was to come in support of the 100,000 roof program, already in progress and running. As a new bill, the EEG, a new name for a new Feed-in Law had to be approved by Parliament. A first proposal was drafted by Fell and eventually introduced for decision by H. Scheer for the SPD and H. J. Fell for the Greens and colleagues of them.

After approval by the Bundestag, the EEG was then turned into law by the "Red-Green Coalition" government. The first FIT by the EEG provided in addition to the provisions of the 100,000-roofs program 45 cents per kWh.

But the EEG faced strong opposition from the EU in Brussels. It took until March 13, 2001, to sort it out. This became then the date for the EEG to become actually effective.

What remains today in essence is that grid access of independent producers is obligatory, electricity from renewable sources has priority in access, the cost-based purchase prices differentiate between different types of power and local conditions, and payments are based on long-term purchase contracts of 20 years or more.

For compensation of the utilities' charges in promoting the FIT, governments fix a compensation fee to be paid by all the clients of the utilities in the country.

For closing the chapter, I should mention the particular role of Hermann Scheer from 1988 when he created the association EUROSOLAR onwards until his too-early demise in 2010. We owe him a great deal for the success of PV and all renewables in Germany and the world. He gave hundreds of talks and speeches, wrote many popular books on renewable energy; some were bestsellers.

A last article, maybe his testament, was published after his demise, in my book Solar Power for the World: What You Wanted to Know about Photovoltaics, published by Jenny Stanford Publishing in 2013 (H. Scheer, "Initiating a Solar Revolution in Germany," pp. 287). Hermann Scheer was my best friend.

Figure 2.17 Hermann Scheer, on the right, with the author in 2009 on the way to the United States, where Hermann received the Karl Böer Award, with the author to give the Laudatio.

2.2.1.5 The FIT opens since the year 2000 the door to PV's worldwide exponential growth

After Japan that kept global leadership of the world PV markets since 1992, Germany took in turn the lead since the early 2000s when the FIT had been adopted there for PV.

Italy and Spain followed Germany in adopting a national FIT for PV.

The first 1 GW of PV was installed in Germany in 2004.

Already in 2009, global PV production passed 10 GW, almost all crystalline silicon modules. Already since the years 2000 much PV was imported from China. For the first time that year cell production costs of $1/Watt were reached through the "economy of scale" and the "learning curve," first effects of over-capacity, and market competition.

Three years later, in 2012, already 100 GW stood installed worldwide; and a lot of it in Germany. The apportionment for the utility clients in Germany reached over €12 billion there in the single year 2010. The government reacted by reducing the FIT payment by 25% in 2010. As the excessive growth of the market went on, the German government reacted by reducing the FIT again by 18% in one year.

As a result, young PV companies lost their profitability. Most of them went bankrupt. It took some years to normalize PV's growth on a more regular path. In 2015 China got ready to take over global PV market leadership from Germany: in 2016 Germany installed 2 GW in the country while China installed 24 GW in 6 months.

By 2018, 500 GW stood installed in total worldwide.

2.2.1.5.1 *2020, PV reached grid parity in China and became economically competitive*

PV was subsidized in China via the FIT until 2018/19. Then after a period of investigation, it was concluded that PV costs had sufficiently come down to make it competitive with electricity from coal-fired power plants. The national government skipped the FIT and all subsidies since 2020.

In the liberalized market that followed PV started then the exponential growth making it the world's No.1 source of power, of clean and cheap power as it were.

Globally the first 1000 GW of total PV capacity was reached under China's leadership in 2022.

2.2.1.5.2 *2024, over 2000 GW installed*

An annual business of hundreds of billions of dollars and euros, PV has become the leading power provider in the world and the cheapest of them all.

Under the lead of China, the world has installed over 500 GW of new PV power capacity in 2024. This brought the total of grid-connected and operating Solar capacity at year-end 2024 to over 2,100 GW.

A fourth of the world's total power capacity, all sources combined, is already PV power in early 2025.

8.3% of all the world's electricity was supplied in 2024 by PV. A major increase over the 5.4% it supplied the previous year. This gives evidence, inter alia, that grid connection problems, including those lack of sufficient grid capacity and curtailment are being progressively solved.

In conclusion, not only in future, PV is already the cheapest source of electricity in virtually all solar climates.

Intermittency is progressively overcome by BESS, the battery storage of various dimensions.

$554 billion (+50%y/y) has been spent in the business of PV that year 2024, most of it as private investments, not from government budgets as is the case for nuclear power.

2.2.1.5.3 *Clean PV together with all other renewable energies is on the spot to preserve the environment and Earth's climate*

Air and water pollution keep progressing in many parts of the world.

Climate change is going ahead; in 2024 Earth's global temperature has already increased by 1.5°C. Wildfires of unseen dimensions before, droughts, and flooding are increasing on all continents as a consequence. The implied costs are enormous.

Even the metro in Mr. Trump's loved city of New York was in 2025 flooded several times by torrential rains...

The renewable energies, with PV power on top, clean and without GHG emissions, are here and now to make the world a better place to live.

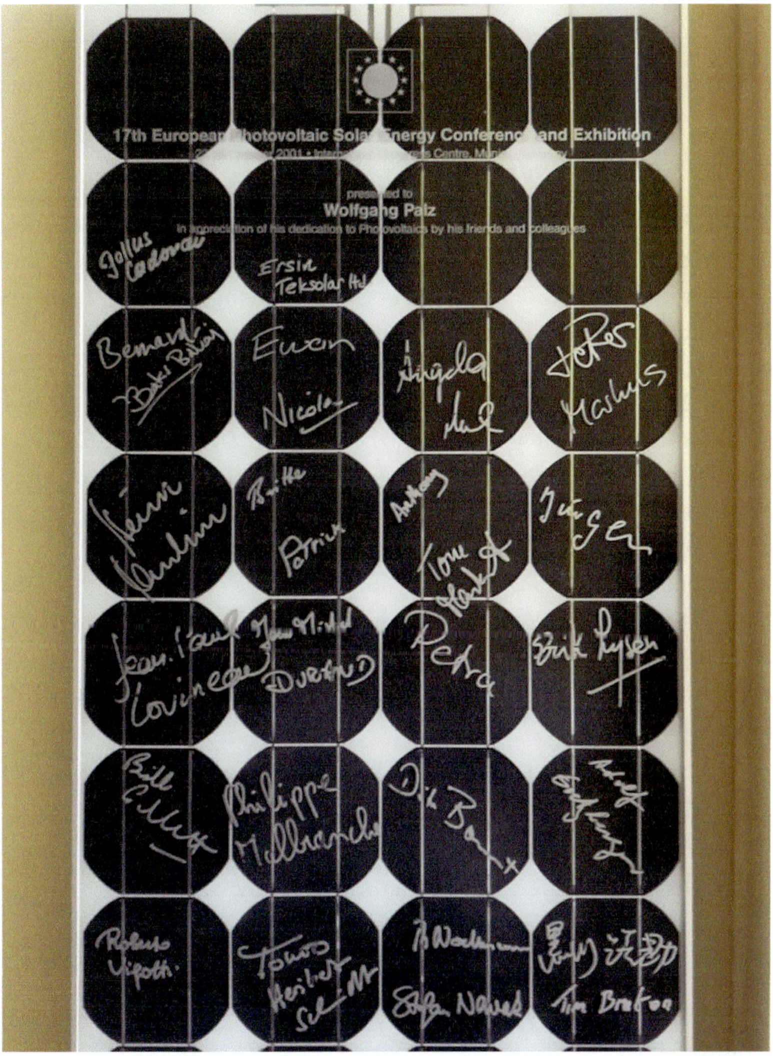

Figure 2.18 PV panel presented to Wolfgang Palz at the 17th EU PVSEC in Munich 2001, dedicated and signed by his friends in recognition of his dedication to PV, organized by Peter Helm.

2.3 Wind Power

The author chaired from the EU Commission in Brussels the European Wind Energy development from 1981 to 1998. In July 1995, he was recognized by the British Wind Energy Association as a "Wind Energy Pioneer."

2.3.1 The Complementarity of Wind Power with Solar Power

2.3.1.1 Wind power today

Wind Power is spreading all over the world. National Energy Administration (NEA) in China, WindEurope, GWEC, IRENA, and other sources found for **2024** the following leading markets (here are the installations in wind power and some electricity generation):

In 2024, globally 121 GW of wind power were newly deployed, bringing the total at year-end 2024 to 1136 GW, or 1.136 TW of wind power capacity. Onshore were 109 GW newly deployed and offshore 8 GW (–2.8% y/y). In 2024 23,000 turbines from 29 providers have been installed worldwide.

China added 79.8 GW, or 68.2% of the world's total, new installations in the year 2024. China's total wind power capacity at year-end 2024 was 521 GW, or 45.8% of the world's total.

The five other leaders of wind power deployment in 2024, after China, were

- Europe (incl. Turkey) +13.8 GW, bringing the total to 251 GW—the EU countries at 231 GW
- The United States +4.058 GW for a total of 154 GW
- Germany +4.022 GW and a total of 73 GW
- India +3.4 GW and a total of 48 GW
- Brazil +3.3 GW and a total of 34 GW

Among the many other countries active in the sector, we just mention:

- UK in 2024 +3.2 GW, 30% of supply; all Renew. 51% of supply

- Spain +3.1 GW in 2024. In 2023, +3.0 GW generating 64 TWh, 23% of supply
- France +2.4 GW in 2024, 26.6 TWh generated; +2.2 GW in 2023, generating 49 TWh, 9.5% of supply
- Canada, Sweden, Turkey, Australia, ...

For 2025, 170 GW of new wind power was installed globally, bringing the total to 1,300 GW, or 1.3 TW, from 1.13 TW in 2024.

In 2025, China is the global leader in wind power development like in all the previous years. In July 2025, 225 GW of new wind power capacity was under construction in various phases in the country, four times the rest of the world combined.

In Europe, following the association WindEurope, 6.8 GW of new capacity was installed in the first half of 2025 bringing the total to 291 GW, 254 GW of it onshore and 37 GW offshore. By year-end 2025, an additional 12 GW is expected.

The EU Installed in first HY 25 5.3 GW of new wind power capacity; by 2030 some 344 GW in total are expected there.

In first HY 2025, the following countries came on top for new installations in Europe:

1. Germany 2.2 GW
2. Spain 0.89 GW
3. UK 0.76 GW
4. Turkey 0.6 GW

By 2030, the world is projected to have 2 TW of wind power running. As they say, it took 23 years to have the first TW in place, and only 7 years after 2023 to get the second TW.

The latest figures for wind power penetration as a share of the total in European countries for 2024 confirm wind power everywhere North-South, East-West:

- Denmark 56%
- Ireland 33%
- Sweden 31%

- Germany 30%
- UK 30%
- Netherlands 29%
- Portugal 28%
- Lithuania 27%
- Spain 25%
- Finland 24%
- Greece 22%

Wind power has come a long way in Europe. In total, Europe had installed in 1990 a meagre 100 MW. Five years later, in 1995, it was already 20-fold with 2000 MW or 2 GW, and the generation of electricity per kW of wind power installed increased from 900 kWh to 1700 kWh.

By 1997, Europe counted a dozen MW-size turbine manufacturers.

Just before the turn of the century, the world had installed in total the first 10 GW in 1998. Now, 25 years later, global wind power capacity increased 100 times!

By 2009, global capacity had reached 160 GW, of which 76 GW was in the EU, 35 GW in the United States, 26 GW each in China and Germany, 19 GW in Spain, and 11 GW in India.

In 2017, 0.5 TW, some 510 GW, had been installed.

At last, in the year 2020, when 749 GW had been reached, wind power's global capacity was for the first time a bit lower than PV's capacity. Since then, the trend of slower wind power growth compared to PV's accelerated; 5 years later, in 2025, the world has already installed twice as much PV than wind power.

2.3.1.2 The complementarity of energy generated from wind and PV

As a rule, wind power generates more electricity in the winter than in the rest of the year. The difference depends on the region considered and lies between 1.5 and 2 of higher generation in winter compared to the summer months.

For China, the difference comes close to a factor of 2, and for Germany it even exceeds the factor of 2.

For PV-generated electricity the contrary is true. On a global level, over some 150% of solar electricity is produced in summer compared to wintertime. As given here above, in Germany, for instance, solar electricity in winter is particularly low. Even more so in recent years as the weather has become rainier.

This seasonal intermittency of the two renewable sources of power actually assures, when seen in combination, some continuous security of electricity supply over the year.

For Europe, more details were already given in our book *Solar Euphoria* (p. 85: "A climatological assessment of balancing effects and shortfall risks of PV and wind energy in Germany and Europe" Kasper et al., *Adv. Sci. Res.* **16**, 119–128, 2019). For the 20 years between 1995 and 2015, the capacity factors of wind power and PV were considered—and both combined. **The complementarity turned out to be excellent.**

2.3.2 Wind Turbine Technologies: The Emergence of "Direct Drive"

Modern wind turbines are horizontal-axis and employ three blades and a fixed-speed induction generator.

Most onshore turbines come today in the power range of 4 MW or so. Offshore ones come 8 MW and up to 16 MW and more.

Two different technologies are used to adapt the rotational speed of the turbine rotor—which is slow, in particular, for large machines—to the rotational speed of the generator. The generator one can be made smaller when using a higher rotational speed, such as generated from the gearbox.

Vestas, the traditional leading Danish manufacturer of wind turbines, employs, as always, gear boxes; for instance, for its 3 MW wind turbine at a rotational speed of 16.7 rpm, a gear box to adapt to a 1500 rpm generator is used, a ratio of 90:1

The Chinese Goldwind, nowadays the world's leading turbine manufacturer, uses "direct drive" turbines instead; with no gearbox but a permanent magnet generator. It may have 40 pairs or more generator poles; the generator's rotor is directly flanged to the rotor hub.

Goldwind uses a license from Vensys in Germany, designed by the "direct drive" pioneer Friedrich Klinger. Klinger has since 1990 worked on the subject in the small region of Saarland. *I know well Prof. Klinger as my home region in Germany is Saarland as well.*

One may also mention here Franz Villinger, a self-made man and pioneer who developed wind turbines in his workshop in Southern Germany. He had noticed that many wind-driven DC generators as battery chargers break down many times; so his idea was to replace them with AC generators. At the 1961 UN Conference in Rome he reported that he had been working since 1956 on a "Small wind electric plant with a **permanent magnet generator." He may have been the first**. His little 500-Watt battery charger, a two-blade-rotor of 5 meters in diameter, pitch regulated and sitting on a telescopic mast, was offered for $1000. The rotor had a high running speed designed for a wind velocity of 8 m/s. He saw a market for remote houses.

2.3.2.1 The challenge: The offshore wind power capacity

As a rule, winds at sea are stronger and steadier than those on land. Offshore wind turbines have virtually no area restrictions like those on land with their multiple kinds of occupation. But offshore turbines are more difficult to install or to float in the sea. **As a rule, by 2023, the installation cost of wind power capacity offshore was still three times that for installation on land.**

- In 2009, the global installation rate was 2 GW offshore and 160 GW onshore, some 1% for offshore
- In 2017, installation rate for offshore with 34 GW in total had a relative growth compared to onshore with some 500 GW, a share of 6.3% of the total
- In 2023, offshore increased even further with 73 GW compared to 944 GW onshore: it came then 7.3% of the total
- In 2024, in Europe offshore came over 10% of onshore: 248 GW onshore and 37 GW offshore. Or 2.6 GW new offshore for 13.8 GW new onshore: 16% of the new capacity came offshore in Europe, when including the UK

At year-end 2024, the world had installed 8 GW of new offshore wind power bringing the total to 80.9 GW, half of it in China, followed by the UK and Germany.

For 2025, an acceleration of the wind offshore market is expected, with 19 GW of new capacity to be installed globally, two-thirds of it in China.

And what about the generation of electricity? That one is bigger offshore, almost by a factor of 2 against onshore: According to the Global Energy Monitor, for 52 GW installed offshore globally, the capacity factor was 41% in 2023 compared to 23% onshore for China or Germany.

In conclusion, offshore wind power is more expensive to install, but produces relatively more electricity.

In Germany, the EU's largest wind power market, the offshore share is higher than the global average: In 2024, in Germany from the total capacity of 72 GW, 63.5 GW stood on land and 9.2 GW offshore. From the total of 150 TWh produced in the year, 20.8 TWh came from turbines in the North Sea and 5 TWh from offshore capacity in the Baltic Sea.

And the world gets prepared for more offshore wind power capacity to come, with monster machines:

- Vestas has been running its V236-15 MW with a 236 m rotor diameter.
- Goldwind has the GWH 252-16 MW with a rotor diameter of 252 m, and has announced still bigger ones to come.
- Siemens Gamesa with a SG DD-276 of 21 MW with a rotor diameter of 276 m, direct drive with permanent magnets. Co-financed with €30 million by the EU, produced in Spain and tested in Denmark.
- Dongfang in China has just presented at year-end 2024 a 26 MW turbine with a 310 m rotor diameter, with its hub 185 m high.
- And some others in China: CSSC Haizhuang or Changhai Electric, and some more...

In a joint statement of the **North Sea Energy Cooperation** from October 28, 2024, the European North Sea countries Belgium,

Denmark, France, Germany, Ireland, Luxembourg, the Netherlands, and Norway demanded: **make the North Sea "Europe's Green Power Plant."**

2.3.3 China's Leadership of Today's Wind Power Industry and the Markets

See also Section 2.2.1.1.1.2, "China's PV exports," for the whole power and electricity picture of China.

Following NEA, in 2024 China installed 79.3 GW of new wind power, more than 60% of global wind power deployment in the year. The total capacity reached 521 GW, 16% of the country's total power capacity and 10% of its electricity generation mix. A year earlier, it installed new 45 GW, increasing its total installation to 441 GW. Since 2013, China has increased its wind power capacity 6-fold. For 2025, the China Electricity Council sees an installation rate of 119 GW of new wind power, bringing the total to 640 GW.

For all the recent years, installation rates for new wind power in China fell just short of around half of the world's wind power markets.

China's offshore wind power market in 2024 came 6.9 GW, bringing the total to 41 GW—this being more than 50% of the total capacity installed worldwide.

The three globally leading turbine manufacturers in 2024 were from China: they were, in order of market share, **Goldwind, Envision,** and **Mingyang.** Goldwind, the number 1, installed 20 GW, 20% more than in 2023, and more than 60% of what it installed in 2022.

Outside China, leading manufacturers were Danish Vestas as the market leader with over 10 GW installed, followed by Siemens Gamesa, from Germany/Spain, and the German Nordex.

2.3.4 The Business, the Employment

The global wind energy business in 2022 was $140 billion for onshore + $34 billion for offshore.

Five years earlier, in 2017, it had been $103 billion and $21 billion, respectively, for half the capacity installed that year, indicating a cost reduction of the wind turbines in the 5-year interval.

The cost of wind-generated electricity on average in China was $0.05 or 5 cents per kWh. That was in 2017. By 2023, it was only half of it at $0.025 per kWh, very low indeed.

The corresponding figures for Germany came with 5 cents/kWh on average in 2023, as high as they had been in China 6 years earlier.

The world counted some 1.5 million jobs in the wind power business in 2023, half of them in China and 300,000 jobs in Europe.

2.3.5 Wind Power: A Long Record of Development

2.3.5.1 The beginnings, the windmills

Man has already used wind for thousands of years for his needs. Leaving alone that, before the arrival of the steam engine, all ships were sailing vessels, they were not rowing boats, and multi wing turbines were in use early on for grain milling, irrigation, and others. It is thought that the first windmills were developed 4000 years ago in Persia and China, vertical-axis ones, it was found.

Several hundred thousand of them have played a pivotal role in the development of rural communities in the US, Latin America, everywhere. Those were horizontal-axis systems.

In the 19th century, 200,000 of those were running in Europe. They disappeared with the beginning of WWI in 1914.

Today's Revival of wind propulsion on commercial ships

In this decade, propulsion of commercial ships by the wind came into fashion again, not only the old-fashioned sailboats, but new technology—in particular, those like the **Flettner rotors from 1924,** using the **Magnus effect.** For the purpose are installed on a ship one or several tubes vertically, cylinders that are in rotation, a must. The rotation is driven by small engines, not the wind. Consequently, the rotating tubes being subject to a wind flow, a force is created on the tube that drives the vessel forward, the Magnus effect.

The **Windship Association** expects over 10,000 ships equipped with them by 2030.

As ships employing the technology are smaller and the service costs are dearer than for conventional freight ships, they are nowadays only used for valuable goods. **The French Arianespace** has used it, for instance, to transport **Ariane 6** to its launching pad in Kourou, South America. Are also shipped on such boats valuable champagne, or chocolate.

2.3.5.2 Denmark, the pioneer of modern wind power

Poul la Cour and his world's first wind-powered electricity generator: Poul la Cour was the world's pioneer for generating electricity from wind power. In 1891, he installed the wind power generator at the school in Askov, the town and the school where he was employed. In this context, one has to remind that Th. Edison started the world's first power plant for New York only in 1882/84. la Cour, on this side of the Atlantic, came really not far behind with his wind electricity plant. By the way, it was a modern-looking horizontal-axis, four-blade rotor. Since 1902 Askov and the villages around got wind power to charge some batteries. The use of the new electricity developed in Denmark, and so did the use of wind power for electricity generation. Many thousand wind chargers from the United States were sold to farmers for some light and radio. By 1918 from the 418 Power stations in the country, a fourth were wind powered. Thousands of wind generators were used for mills and irrigation on the farms, but later in the 1940s, only 25 of those were left.

SEAS, the Danish utility, going ahead since 1947: Since 1947, SEAS investigated small wind turbines on their grids. Then wind measurements were performed at many places with anemometers, in cooperation with the Danish Met Office. Thirty different blade designs were tested. Then a pilot mill with a rotor diameter of 7.5 meters was built in 1950 near **Vester Egesborg** in Zealand. There, experiments were performed on the effects of various blade tip speeds under different wind velocities. This pilot did not have a tower yet and was held with sales. It was running until 1960 when it was dismantled.

The next step was the turbine on the island of Bogoe. There, SEAS reconstructed an old DC turbine from 1942, from the firm Smith & Co into an AC turbine. From two-blade it was made three-blade with a swept area of 132 m^2. It fed a 45 kW AC generator. The turbine was running for many years.

Next step was that, on proposal by SEAS, the Association of Danish Electricity Works DEF appointed a Wind Power Committee to carry on.

Gedser: On application by the committee, the Danish government granted DEF a budget (some €41,000) for the erection of a larger machine near Gedser. Like the Bogoe machine, it was one of those DC machines from 1942, enlarged to a three-blade rotor with a swept area of 450 m^2 sitting on a 25 m tower. At 15 m/s it delivered 200 kW. The blades had been optimized by wind tunnel measurements by SEAS. It was stall controlled. **J. Juul** was the engineer in charge of Gedser. Gedser was completed in summer 1957. After adjustments, it stayed since 1958 connected on the SEAS grid.

Gedser was a kind of grandmother of modern wind turbines.

NIBE: After the oil crisis in 1973, the utilities Elkraft and Elsam formed a partnership with the Danish Ministry of Energy on a wind power program. Thereupon, two 630 kW machines were put into operation in 1979/80 at NIBE in Northern Denmark. Later, Elkraft built five turbines, each of 750 kW, the Masnedoe Windfarm. Danish Wind Technology A/S supplied the machines.

Tvind: A wind turbine was built by enthusiasts, not experts, in 1975 by local people in the town of Tvind. It was a grassroots movement triggered by the anti-nuclear movement in the country. Teachers and students built there for their school, with their own money, a **2 MW wind turbine** from scratch. It was completed after 3 years in 1978. 27 m-long glass-fiber-reinforced plastic blades—à la Hütter—they did build themselves for the purpose. **Tvind was Europe's first MW-size wind turbine.** In 1988, **Ove Dietrich**, the Chief Executive of SEAS declared in public that at some sites Danish wind power had become economically competitive by that time.

Markets take off: From 1982 to 1986, thousands of Danish wind turbines were sold to California, a market promoted by local tax credits. In 1988 1630 small-scale wind turbines with a total capacity of 110 MW were grid connected in Denmark.

2.3.5.3 US study: two-blade or three-blade rotors?

Jakobs at Fort Myers in Florida built several thousand small machines as battery chargers.

They investigated extensively rotors with different numbers of blades, in particular two-blade and three-blade rotors. They eventually went for the three-bladers, the others showed excessive vibrations.

Jakobs sold thousands of 3 kW machines, also in Europe, between 1931 and 1957. They were pitch controlled. Meyers gave a 5-year guarantee. They came $1000 and had no maintenance cost, they said.

Jakobs sold also hundreds of them, charging a 21 kW battery, for cathodic protection of steel pipes.

2.3.5.4 Germany: The beginnings and what followed

Honnef: Hermann Honnef had patented since 1931 his "GrosseWindKraftwerke." Knowing that wind flows stronger and steadier at higher altitudes above ground, he thought one had to build very high wind towers. He was ahead of his time: he proposed turbines with over 430 meters high towers, and up to five rotors attached to one single tower. Each rotor was supposed to be 160 meter in diameter. He couldn't build any as he was jailed by the SS in 1934. It took until today that such high and powerful turbines were eventually built in the 2020s, simpler and of conventional design of one rotor per tower, for offshore, as it were.

U. Hütter: In December 1949, the Swabian utility in Germany proposed a committee for the study of economic viability of wind power. Then followed a design competition for a 100 kW plant at 8 m/s. Allgäuer Tool Prod. Corp's offer, prepared by Hütter working then for that company, was selected. Followed in 1955 the establishment of a Wind Development Group. Many utilities and other firms participated in it.

Hütter's StGW-34 was built in 1955–57 in Stuttgart, installed in Stötten. It was first a two-bladed machine, as U. Hütter insisted that three blades were one too many. Until over the years too many problems had appeared with it and it was converted to a three-blade machine under the same name. It had then a 17 m diameter rotor and was pitch controlled. Some 200 of these machines were built and sold. They are said to have been until the 1980s among the most successful commercial machines. Hütter saw since the 1980s wind power on the verge of mass production in the industrialized countries.

GROWIAN: Referent Klein, in charge of wind power development at the Ministry BMFT in Bonn also wanted it big, a new record. The rotor had to be over 100 meters in diameter, if just a little bit. It had to be Germany's first MW turbine.

Planned since 1978, the "Grosse Wind Anlage GROWIAN" was built by the company MAN and financed by the BMFT. It was a two-bladed 3 MW machine. It was running since 1983 no more than 420 hours in total and then over. Inter alia much too heavy. In 1987, the plant was dismantled.

Monopteros: The company MBB built, financed by the BMFT, the one-bladed Monopteros, in different sizes. The M20 had 25 kW of power, then the M30 built with the Italian Riva Calzoni since 1983 was a 200 kW machine. Then several M50 of 640 kW were built since 1989. It had on a tower 60 meters high a one-blade rotor of 56 meters. In 1992, all machines were dismantled. No success! **Later MBB participated in a 3 MW two-blader Aeolus II built with the Swedish Kvaerner.**

Ad-hoc Ausschuss Grosswindanlagen: After the GROWIAN disaster, the Minister in charge of BMFT called in 1991 a committee on large turbines. It followed actually a study on large wind turbines that we had conducted with wind power experts at my own wind program at the EU Commission in Brussels. The report of that study was already available in 1990.

As an EU official and the leader of the Commission study on the same subject, I was invited by the Minister, next to the German officials and experts, Mrs. Sandtner, Kleinkauf, Hau, Molly and Wobben. The German committee recommended in its report by 1992 an aggressive development program. But there was no follow-up.

Now in 2025/26, Germany comes on top of the wind power installed in all European countries. Its companies Siemens Gamesa and Nordex come, behind the Danish Vestas, among Europe's leading turbine producers.

2.3.5.5 Sweden

The utility Vattenfall investigated wind power since the 1970s.

The turbine Aeolus I, a two-blader built by the Swedish Kvaerner with MBB, was installed in Gotland at a place where by now hundreds of machines are installed.

Aeolus II of 3 MW, also a two-blader built by the same consortium, was running from 1993 until 2008 on Jade Windpark in Cuxhaven, Germany.

Another two-blader of 3 MW was **Näsudden II,** also built by Kvaerner, which was in operation since 1993 and till 2007 in Gotland.

2.3.5.6 EU Commission: The first industrial MW machines in Europe WEGA I and WEGA II

In the Commission's program that I directed since 1981, we felt concerned by the issue of MW-size turbines. We saw the need for the larger markets of the future, and a standstill since the GROWIAN disaster. My attention to the limitations of enlarging the turbine sizes was drawn first by David Milborrow from the UK utility CEGB. As the power of a turbine obviously increases with the swept area, or the square of the rotor diameter, the machine weight and its cost increase with the cube of that diameter. To avoid excessive mass or weight of a machine, there ought to be somewhere limits for turbines of reasonable size and cost.

Therefore, we commissioned a group of leading European wind power experts to analyze this question in detail. The result was that because of the physical limitations, there is an optimal size for cost-effective machines. There is an optimum of 2 to 3 MW, but it is very flat and depends on the particular design of a machine. For offshore turbines, it is larger.

In conclusion, I decided to initiate an experimental international program on MW-size wind turbines in Europe. We called it the WEGA program, where WEGA stands for the German "WindEnergieGrossAnlagen." We implemented two WEGA programs, one more experimental, and a follow-up WEGA II more market oriented.

WEGA I was started in 1985 and three machines were built and tested, one in the UK, one in Spain and one in Denmark. They all started running in 1989. Conventional electric utilities were involved in all three. The three together served as a joint European development platform with intense communication and consultation among the stakeholders involved. The EU Commission bore part of the cost of constructions and experimentations, €7 million, as it were.

Identical measuring devices and a joint reporting software were used. Data acquisition was made compatible. Independent contractors were commissioned from Brussels to coordinate the measurement program and to evaluate the results.

The Danish machine was installed at Esbjerg in Jutland. It had a power rating of 2 MW and a 60-meter rotor diameter. It was an extension of the design of the NIBE turbine. After some years of experimentation, ELSAM, the owner, had had enough of it and blew it up.

The Spanish machine, actually a German one from MAN, in cooperation with the utility Union Electrica Fenosa, was installed at Cabo Villano at the North Western corner of the Iberian peninsula, near Santiago di Compostela.. Our machine was the first at this windy place; nowadays there are hundreds of them of all kinds of sizes. The rotor diameter of this AWEC 60 was also 60 meters like the Danish one.

The British machine was installed near Richborough in Kent near the sea. The machine actually neighbored a conventional power plant. The rotor diameter was 55 meters and the particularity was that the blades were made of wood/epoxy composite. Since the tower was 45 meters high, the owner of the plant, the utility POWERGEN, said that the whole plant was higher than London's

Nelson Column, which is 52 m high. The company Howden from Glasgow was responsible for the design of the machine. Besides the usual engineering tests, that were part of WEGA I, the testing program also included interface issues such as bird life, noise and TV reception.

The Howden Machine was the lightest of the three machines and so the most cost effective.

WEGA II: WEGA I machines were experimental and allowed us to achieve the basic know-how of MW-size machines; they did not have a strong industrial component. This was then the purpose of WEGA II. The idea was to get involved the leading European wind power industry that was already in the market. That industry should get a push with WEGA II to implement such larger and more competitive machines.

We published from the Commission a call for proposals for WEGA II in 1991. The program extended until 1997. We were successful in attracting some of the European companies that are even leaders up today. Some of the participants were:

- Vestas, Demark
- Bonus, Denmark, later bought by Siemens
- NedWind, the Netherlands
- Enercon, Germany

The rated power of these machines was 1 MW. The rotor diameter varied from machine to machine between 50 and 55 meters. All these machines showed eventually considerable improvements compared to the first generation of machines of WEGA I.

The weight of the blades and the nacelle were only fractions of those machines.

The total cost of WEGA II was €25 million, of which we from the Commission contributed €7 million.

The benefits of this program were tremendous. It could be demonstrated that manufacturing costs could be reduced to €1 million/MW.

An extensive evaluation of WEGA II and tests were performed and published by the Commission's Offices in Luxembourg in the year 2000.

2.3.5.7 The United States

As early as 1888, a somewhat exotic **C. Brush** developed a wind-powered battery charger with a DC dynamo, for the needs of his own house in Cleveland, Ohio. Its tower was 18 meter high and the rotor was of those traditional multi-blade ones, of 17 meter diameter. It was one of the world's first wind power plant, but it had no follow-up.

In the 1940s, **P. Putnam** designed a 1.25 MW turbine in Vermont. It was built by the company Morgan Smith and was in total only running some 1000 hours. It was wartime, and there was no follow-up either.

Development of wind power started in earnest in the United States in 1975 with the experimental MOD machines, triggered by the first oil crisis. It was a government program, managed by the energy Department ERDA of the Federal Government together with NASA (the Lewis Research Center in Cleveland), and various industries, depending on the project, GE, Westinghouse, United Technologies, Boeing. It ran between the late 1970s until the 1990s and cost $330 million:

- MOD 0: 1975–1982, **100 kW**
- MOD 0A: 1977–1984, **200 kW**
- **MOD 1: 1979–1981, 2 MW, 61 m diameter: the world's first industrial MW wind machine**
- 3x MOD 2: 1982–1988, **2.5 MW** with Boeing
- MOD 4: 1982–1994, **4 MW** with UT
- MOD 5B: 1987–1996, **3.2 MW**, 97.5 m diameter, with Boeing

Early on the MOD managers had studied the European machines Gedser and that from Hütter.

They adopted Hütter's wrong design philosophy: two-bladers with one blade less than three-bladers are lighter and hence cheaper. That also proved wrong with the MOD machines that were all two-bladers and none of them had a commercial follow-up.

The US GE Vernova is these days one of the world's leading turbine manufacturers. So far, it has sold over 57,000 machines and employs 75,000 people. At its origin are also Ecotecnia,

Alstom, Enron, Zond, the German Tacke that were all bought in the late 1990s.

2.3.5.8 The emergence of China's market

A certain **Wu Yundong of the Zhejiang Research Institute of Mechanical Science** had developed in 1983 the ZFD 13, a three-blade 20 kW machine. It had a rotor diameter of 13.4 meters and a rotational speed of 66 rpm. It weighed 5 tons, half for the tower. One pilot turbine was operated since 1985 on the Dachen island with a diesel set where it survived two typhoons. The studies showed that wind power was at that time already economical in China.

In our book *Wind Power for the World* (2014), also published by Jenny Stanford, **He Dexin**, the Chairman of the Chinese Wind Energy Association, had given an overview of how wind power development started in China. In the early days, in the middle of the 1980s when nobody had produced MW turbines yet, to start with wind power, China had first co-operations with Denmark and other European countries.

In 2012, the total capacity in China reached 61 GW of grid-connected capacity, and 12 GW off-grid. In 1996, the total had only been 0.057 GW or 57 MW.

In 2007, China started offshore wind farm development with a 1.5 MW direct drive turbine from Goldwind. By year-end 2012, already 390 MW offshore had been reached.

2.3.6 Future Outlook for Wind Power

The United States got with Trump a new president who made it clear that he will do the utmost to stop all wind power development in the country.

In a Memorandum of January 20, 2025, the Trump Administration ordered a Moratorium on New or Renewed Federal Actions for Onshore and Offshore Wind Projects, Pending Federal Review.

But the president's executive orders can also be a bit contradictory. After all the despair since his first announcements on energy, **here comes the good news**. In May 2025, the White House green-lighted resuming work on a large offshore wind power park off the coast of New York: 810 MW with 54 turbines at a cost of $2.5 billion.

And there is China. Its commitment and leadership in the wind business and its worldwide engagement enlightens the global outlook.

And there are, in particular, its links with wind power development in Europe:

In Europe, the EU Commission is firing year after year its support measures in favor of wind power development.

In October 2023, the Commission set out an immediate plan of actions for Europe's wind power industry, which demands:

- A doubling of wind power capacity up to 500 GW by 2030;
- Acceleration of permitting proceedings for turbine installation;
- Increase by a factor of 10 the offshore installations to 12 GW for every year to come.

Followed in August 2024 the European Wind Power Package, which demands:

- Higher investments in the wind power supply chain, such as the turbine factories;
- Offshore foundation factories;
- Grid equipment factories;
- Cable factories.

No doubt, the EU will do its best to accelerate wind power uptake in Europe...

Following the Association "Wind Europe" +140 GW of new wind power is expected in the EU by 2030, in addition to the 231 GW installed by year-end 2024.

At its annual Conference of April 2025 in Copenhagen, Wind Europe issued a "**Call for Action.**" It referred to the "EU's Industrial Deal" that puts electrification and the renewables at the heart of its industrial future. Currently, wind electricity meets 20% of demand and by 2030 the EU aims at reaching 35%. It is essential, it says, that the EU permitting rules are applied everywhere like in Germany that permits now already seven times more onshore wind than 5 years ago.

2.4 Other Clean Energies, Electricity Storage, and Applications

2.4.1 Solar Thermal Conversion

In 2023, some 800 million m^2, or 800 km^2, of solar heat collectors were installed in total worldwide. The installation rate increased all over the years: In the year 2000, it had been only 88 million m^2.

The corresponding energy in 2023 did correspond to 491 Mtoe and 148 Mt of GHG emissions were avoided this way in the year.

For 2024, the IEA SHC program found 777 million m^2 globally installed, less than the figures here above from INTEC; those seem to be more reliable.

The sector employed some 345,000 people. Global business outside China reached $30 billion in 2024.

As part of the IEA Heating and Cooling Programme, a very detailed analysis had been performed for the year 2022, the "Solar Heat World Edition 2024" by the government of Austria.

In an extraordinary effort involving dozens of personal contacts, the authors Wolfgang Gruber Glatzl and René Baumann from INTEC, Austria, and Bärbel Epp in Germany were able to present a very informative overview of the global state of the art of the solar heat collector business. Here is a summary (areas in km^2):

Year 2022: World total 775 km^2, + 21.5 added in the year, total corresponding to 47 Mtoe and 150 Mt of avoided GHG emissions:

- China 566.3 km^2
- Turkey 27.4 km^2
- US 26.3 km^2
- Germany 22.6 km^2
- Brazil 22.1 km^2
- India 19.6 km^2
- Australia 9.5 km^2
- Mexico 5.9 km^2
- Greece 5.4 km^2

- Italy 5.3 km^2
- Spain 5 km^2
- Austria 4.6 km^2
- Israel 5 km^2

One hundred eleven million single-family houses had SHW year-end 2022:

- China 75 million
- India 8.7 million
- Turkey 6.3 million
- Brazil 5.7 million
- Germany 1.8 million
- Greece 1.4 million
- Austria 1 million

Six million multi-family houses had SHW, too.

Three hundred sixty-six towns and cities had solar district heating.

Figure 2.19 View of Chinese solar heat collectors of the evacuated tube-type.

The technologies:
Globally

- 69% of heat collectors were of the sophisticated evacuated tube type, mostly so in China.
- 25% were flat-plate collectors.
- 6% were unglazed.

In Europe

- 79% were flat-plate collectors.
- 19% were evacuated tube collectors.
- 2% were unglazed.

2.4.2 Bio-Energy

With information from the "Global Bio-energy Statistics Report 2024," the World Bio-energy Association in Stockholm, Sweden.

Bio-energy provides 9% of the global energy supply in 2023. With some 700 TWh it generates 8% of global electricity.

2.4.2.1 Liquid bio-fuels for transport, carbon neutral fuels

In 2023, some 180 billion liters of bio-fuels were generated and consumed worldwide.

The global market is worth $141 billion in 2025.

Some 2.8 million jobs were involved, 1 million of them in Brazil.

Bio-ethanol

The world produced in 2023 116 billion liters of bio-ethanol. It increases the octane value of gasoline when blended with it. In the United States and Europe, it is called E10 when blended with 10% of alcohol.

The first producer in 2023 was the United States, with 59 billion liters or 15.8 billion gallon. Since the year 2000, the annual production volume increased 10 times in the country.

The US Energy Information Administration (EIA) anticipates for 2025 a production of 2.9 million barrels per day from 2.8 barrels per day in 2024. It may increase further to 3.1 million barrels per day. All this growth should go to export.

The United States exports 1.3 billion gallons to 87 countries.

The share of ethanol in all US gasoline is 3% without further notice.

Since 2008, the Environmental Protection Agency (EPA) set the Renewable Fuel Standard (RFS), giving the volume requirement for each fuel category such as ethanol, which are increased every year.

In the United States, ethanol is produced from corn kernel starch.

The E10 blend is called gasohol.

Brazil is the world's second biggest producer of fuel ethanol.

Brazil was a pioneer, starting the fuel ethanol with the Pro-Álcool plan in 1975. Development accelerated in 2003 with the introduction of "flex-fuel" vehicles by VW, Ford, and other car manufacturers that can run with gasoline of any blending rate.

With 35 billion liters, the country produces today half the volume of annual US production and a fourth of global production. Brazilian alcohol generation is linked to that of sugar as both are products of cane, mostly harvested in the southern part of the country, the area of Sao Paulo and neighboring ones on 10 million hectares of land.

Fuel ethanol makes 34% of energy consumption in transport.

In 2024, hydrous ethanol, the E100 came 20 million m^3.

Brazil has a mandatory ethanol blend. In October 2024 President Lula signed a new "Fuel of the Future" program. The purpose is to increase the blending rate of anhydrous alcohol from 18% to 27% now to 22% up to 35%. Experts see soon coming E30 in the country.

Some ethanol is exported, mainly to South Korea.

5.5% of Brazil's electricity is produced from sugar cane residues.

Other countries and regions generating and using fuel ethanol are India with 6.4 billion liters, third in order of market volume, then the EU and China.

Raw materials used for conversion are maize, wheat, and sugar beet.

Europe generated in 2024 5.38 billion liters of bio-alcohol.

The base is the EU Directive RED II from 2018, with 14% of energy consumed on road and rail by 2030 to be renewable.

BBD: Biomass-Based Diesel, Bio-diesel fuels

Global total some 66 billion liters/y

Fatty acid methyl ester (FAME) production in 2023 was some 50 billion liters. Indonesia led with 14 billion liters produced from palm oil, followed by the EU with 13 billion liters, from rapeseed, used cooking oil, and then Brazil with 8 billion liters produced from soybeans.

Renewable diesel HVO produced by hydrogenation of bio-residues and wastes came some 16 billion liters: in 2023 the US with 11 billion liters, the EU 4 billion liters and China 1.4 billion liters.

2.4.2.2 Biogas

Anaerobic digestion of biomass, i.e., in the absence of air and oxygen, generates biogas, a mixture of the energy carrier methane and CO_2. In the presence of air, compost is generated instead.

An infinite number of small microbes is driving the process.

Concerned are in particular the organic wastes that are eliminated that way. And there are gigantic quantities of those produced inter alia by all animals, and the 8 billion humans as well. Life on Earth would have been impossible without the permanent cleaning of all that mess by the little biogas producers. Billions of those microbes can hide in just a cubic cm of material.

The subject is of great importance and for details, the reader has at his disposal a big specialized literature.

A well-known case of modern biogas generation on a large scale is Germany. Organic wastes are treated in combination with fresh biomass crops. Germany has some 10,000 of those digesters running. The biogas generated is employed to run 5.6 GW of power capacity, in principle enough to meet the needs of 9 million households. The small combined heat and power plants (CHP) fed by the methane content of the biogas provide in addition the heat for 2.5 million households. It makes 10% of all renewable heat used in the country.

1.6 million hectares, 9% of Germany's agricultural land, is used to grow biomass raw material, mostly maize, for mixing in biogasifiers.

In China, biomass waste gasification on a family scale has been used for a long time. Maybe some 100,000 of them are still in operation.

The EU has today some 50% market share of the global biogas business that is said to have amounted to $83 billion in 2024. Of concern is in particular OFMSW, the anaerobic digestion of Organic Municipal Solid Waste.

Following the European Biogas Association, the EU had in 2023 treated some 22 billion m^3 of it. After cleaning the raw biogas, there remains enough methane to provide 7% of the EU's natural gas consumption. It is used for transport, in buildings, in the industry and for power generation.

Landfills

Cities like to get rid of all their residues and wastes in landfills. Globally, there are many of them: 48,000, most of them in the industrialized countries, the United States, Germany, etc. They handle 2 billion tons of municipal solid waste per year.

Those landfills produce and release methane, a lot of it. The US EPA estimates that the yearly release of "waste methane" is equivalent to carbon emissions from 23 million cars.

2.4.2.3 Pellets

Some 50 million tons of wood and woody pellets were produced and consumed globally in 2023—52% of them in Europe, 32% in North America, and the rest in Vietnam, China and other countries. Their commercial value amounted to some $15 billion in 2025.

In Europe, pellets are used for 34% in industry; for electricity generation, CHP. Sixty-six percent are employed for heating homes and for commercial purposes.

Heating with pellets is cleaner than burning wood in wood stoves. It may be used instead of heating oil, and it may be somewhat cheaper.

Germany had in 2022 some 650,000 pellet home-heating systems.

2.4.2.4 Heat and Power, CHP

It is said that in 2022 the world employed some 96% biomass for the **renewable heat** it uses.

For the 8% of **renewable electricity** mentioned earlier, the corresponding capacity of bio-power was 150 GW in 2023, led by

- China, 31 GW
- followed by Brazil, 18 GW
- the United States, 11 GW
- India, 11 GW
- Germany, 10 GW

Accordingly, the electricity produced in 2023 was then for

- China, 130 TWh
- the United States, 57 TWh
- Brazil, 55 TWh
- Germany, 48 TWh
- Japan, 37 TWh

Combined heat and power plants recoup the heat that comes obligatorily when electricity is generated. As mentioned above, CHP is often used for biogas combustion.

It is well employed as well with solid biomass, units of up to 30 MW or so. A combined efficiency of some 75% can be reached, 25% for the electricity generation.

For the purpose, the biomass is fired in a high-pressure steam boiler, or alternatively gasified. For the engine, a turbine or a combustion engine is used. For micro-CHP, a fuel cell may also be used when gas is available.

2.4.3 Green Hydrogen and Other Green Fuels Derived from Green Electricity

This book is a picture of a world in transition, a global energy change. Enormous financial and social resources are being mobilized.

So far two things have come to the forefront:

1. There is a worldwide revolution in the electricity sector going on. The clean renewable sources of electricity,

with PV and wind on top, dominate the global power markets today. Thanks to the new mass markets that they developed since the turn of the century, electricity generated from PV and wind power has become the cheapest today, cheaper than electricity from coal and all fossil fuel-fired plants, and even more so than that produced by new atomic power plants. What a success !!!

2. But climate change is accelerating together with the terrible disasters that come with it. CO_2 and all GHG emissions stabilize instead of going to zero. Until 2024, they even increased from year to year as the use of fossil fuels globally grew instead of declining.

Conclusion: After the power revolution, the world needs a clean fuel revolution.

The clean fuel that is abundant in nature is hydrogen, a component of all the water that is there. But the hydrogen of choice is not the grey hydrogen generated from water via fossil fuels, already produced and used today in large markets, nor is it the pink hydrogen involving nuclear energy–powered electricity. It is green hydrogen and its derivatives are ammonia and methanol. Hydrogen is the lightest gas that exists. The whole Universe was formed out of it. It is volatile and inflammable (to form water), while the derivatives are liquids that can be easily used.

The only ingredients of green hydrogen are water and the electricity derived from the renewables, such as PV and wind power, feeding an electrolysis device.

As long as that kind of electricity was expensive, not surprisingly the green hydrogen remained expensive as well. This is no more the case today. As was the case for PV in the past, today's costs of the green hydrogen will come further down by mass production and use.

Details about green hydrogen and its future perspectives can be found in *Hydrogen Insights 2024*, published by the Hydrogen Council in Brussels.

Today, about 1 million tons/year of green hydrogen are produced, 1% of the grey hydrogen market. In May 2024 global electrolysis capacity for it was 1.75 GW, most of it in China.

7 to 11 million tons/year of clean hydrogen from renewables have a high probability to come online by 2030 following this report.

It could become a lot more: 1500 projects involving $680 billion have been announced. On the horizon of 2030, 4.6 million tons/year capacity passed the Final Investment Decision (FID) stage. $75 billion are committed for green hydrogen plants by 2030 (China $31 billion).

The electrolysis capacity for green hydrogen up to 2030 is probably for 90 GW, 26 GW already have an FID.

As is already the case for all renewable energies, development of green hydrogen in China outpaces the rest of the world.

2.4.4 Hydro Power

While green hydrogen, discussed above is, one may say, the youngest of the world's renewable portfolio, hydro is the conventional among the renewable sources of power. In 2023 hydroelectricity provided some 15% of global electricity generation and consumption, in South America even 45%.

Following the World Hydro Power Outlook from IHA, the International Hydropower Association, the world had installed in 2023

- 1416 GW (+13 GW in the year—half of it in China)
- 4185 TWh of electricity generated from it
- 182 GW (+6.5 GW in 2023) of total capacity is pumped storage

The association estimates that globally there remains a potential of some 2000 GW for hydropower.

500 GW of them in the "big 100" pipeline are projects to be decided yet.

The biggest to have been announced is the Yarlung Zangbo Hydrodam in Tibet/China. The river flows into India, where it is called the Brahmaputra.

With a generation capacity of 300 TWh a year, it would be three times the output of the "Three Gorges Dam" in China. It will be enormous. In comparison, the famous Hoover Dam in the United States generates "just" 4.2 TWh.

For the time being, the river runs naturally through a 6 km deep canyon, the deepest in the world, and longer than the Grand Canyon in the United States. In comparison, the Grand Canyon is "only" 1.5 km deep.

There is no timeline set as yet for this monstrous project; its cost has been evaluated at $137 billion. It is part of China's 14th 5-year Plan.

2.4.5 Geothermal Energy (Power and Heat, Heat Pumps GHP)

Unlike all the energies presented so far in this book, geothermal energy is not a derivative of solar radiation on Earth. It is fed by our planet's own heat resource—7000°C in the center, fixed in billions of years by slow nuclear decomposition of uranium. Volcanoes and earthquakes tend to demonstrate the virulence of this energy.

Following **"Geothermal Rising,"** the Association from 1972, and the later founded IGA, the International Geothermal Energy Association, the world made use in 2023 of Geothermal heat in 88 countries.

Geothermal Power was run in 31 countries, with the leading markets in the following ones:

- The United States, 3.7 GW
- Indonesia, 2.4 GW
- The Philippines, 2 GW
- Turkey, 1.7 GW
- New Zealand, 1 GW
- Kenya, 1 GW
- Mexico, 1 GW
- Italy, 0.9 GW

In total, the world had in 2023 a Geothermal Power capacity of 16 GW, running in 671 power units. Technologies employed were "flash" cycles, dry steam, and organic cycles.

For heating, in total 173 GW_{th} of geothermal heat was in use, 152 GW for heating and cooling, and the rest for health, tourism, and others.

Ground source heat pumps (GHP) employ the heat, just 1 or 2 meters under the ground surface. Their temperature remains constant, if everything goes well, between 4.5°C and 21°C year round. One uses shallow horizontal heat pipes or deeper vertical heat collectors in closed-loop systems.

GHP heat pumps come with a higher upfront investment than air source heat pumps, but they are quieter, more efficient, and last longer.

An example is Paris in France, which is sitting on a big bubble of hot water, the Dogger, 1600 to 1800 meters deep. It has been already exploited since 2010. Nowadays, it supplies heating to some 60,000 dwellings. The investment cost of up to €120 million per project is not negligible, but 100,000 tons of CO_2 emissions are thus avoided every year.

2.4.6 Marine Energies (Wave and Tidal Power)

2.4.6.1 Wave power

There is a lot of hope in wave power as viewing its strength looks tremendous. *I experienced myself pressure from politicians in the EU Parliament to do something about it when I was in charge of energy development at the EU.*

In practice, wave power is difficult to convert into power that is usable. In particular, since climate change made the seas more virulent: now there roll the "century monster waves" every year.

In reality, today there are no commercial wave power plants running.

2.4.6.2 Tidal power, ocean currents

Tidal power plants of a few 100 MW are in operation and running.

The oldest of 240 MW capacity from 1966 is at the Rance in France.

The latest under construction, the MeyGen in Scotland, will have a power of 400 MW.

There are a lot of water currents in the seas. The great Gulf Stream system is one of them.

A famous example is the one at the Messina Strait between Sicily and the mainland. In *Odyssey*, the poet Homer made a story out of it. In reality, it is a tidal current that inverts direction every 6 hours.

2.4.7 Electricity Storage (Pumped Hydro, Compressed Air, Electrochemical Storage)

2.4.7.1 Pumped hydro energy storage

As mentioned in a previous section, the world was running in 2023 some 182 GW of pumped hydro.

2.4.7.2 Compressed air energy storage (CAES)

There are only a few CAES running today. The oldest is grid connected since 1978 in Huntorf in Northern Germany.

In the United States, there is the McIntosh plant.

One of the more recent ones is the one in Hebei, China, of 300 MW capacity, operational since 2024. They are all Salt Cavern Compressed Air plants.

The one in Jintan in Eastern China broke ground in late 2024. It will be the world's largest when finished:

- Two 350 MW units
- 2.8 GWh storage per charge
- 1.2 million m^3 gas capacity in total

2.4.7.3 Electrochemical energy storage

Following the Volta Foundation, the international network of battery professionals, the world had installed at year-end 2024 **160 GW in total, a capacity of 363 GWh of electricity supply. This was 10 times more than 5 years earlier! In 2024, 70 GW of storage was added globally.**

China had at that time 74 GW of battery storage installed, +50% y/y. For end of 2025, 100 GW is expected in the country and by 2030 even 200 GW in total.

By 2030, a further strong increase up to global battery capacity, of 1200 GW of all kind of sizes, is expected.

The cost of battery storage is evaluated some €200 the MWh by McKinsey. In comparison, the cost of stabilizing the net by gas power plants can be as high as €1000/MWh or even €3000/MWh.

The cost of the batteries on the international markets came down from $800/kWh to just $115/kWh in 2024, triggered by the mass market in China and lower cost of lithium..

Falling cost of lithium, iron, phosphorous (LFP) batteries makes them the darling of the markets. The more conventional one, lithium, nickel, manganese, cobalt (NWC), are losing market shares down to a little 1% by 2030, as experts do expect.

2.5 China, Renewable Energies' Wonderland

2.5.1 China, Top of the World: Recent Achievements

2.5.1.1 2024

Following the Chinese National Energy Administration, the country in 2024 with 1407 GW, or 1.407 TW, of PV+wind power capacity coming close to its thermal power capacity of 1.444 GW.

Wind power and solar PV made up 42% of the total installed power capacity in the country.

In the year, China added 277 GW of **PV** capacity, bringing the total to 887 GW:

- **Wind power** increased by 80 GW attaining 521 GW in total,
- **Hydro** became 436 GW in total,
- **Nuclear** 61 GW,
- **Power capacity of new coal-fired power plants** increased by 30 GW.

In terms of electricity generation, the country's share of PV increased by 28%, that from wind power by 11%, and that from hydro as well be 11%.

Thermal power supplied only 1.1% more in the year.

China Electricity Council (CEC) and others released the Electrochemical Energy Storage Station Industry Statistics 2024. In the year, 62 GW of batteries have been put into operation, 41% on generation and 57% on the grid. 67% use a 2-hour storage duration. 99.9% use lithium, iron, phosphate. Large-scale stations of over 100 MW dominate the market.

2.5.1.2 Q1 2025

The country added 74.33 GW of new PV and wind power, bringing their total to 1.482 GW. Total PV capacity reached 946 GW, and that of wind power 536 GW.

Thermal power, now at 1.451 GW, was for the first time lower than PV+wind power.

All the renewables' power in China reached 1.966 GW end of March 2025, some 57.3% of the total installed capacity. PV (946 GW), wind power (535 GW), hydro (438 GW), and bio-power (a small 0.046 GW).

$13 billion were invested in the power grids.

For electricity generation, solar PV + wind power supplied 22.5% of the national total, +4.3% y/y. Their share is steadily increasing, from 3.9% share 10 years ago to 9.4% 5 years later in 2020.

Together with hydro, 33% of total supply in Q1 2025:

- Wind power, 13%
- PV, 10%
- Hydro, 10%

Together with the larger use of renewable energies, China's CO_2 emissions are decreasing—In Q1 2025 by 1.6%.

2.5.2 New Policies

Six Chinese Ministries declared in early 2025 a joint Heat Pump development action plan. It strives to increase energy efficiency of key heat pump products and promote innovation.

New promotion of Virtual Power Plants (VPPs): VPPs are a new model aggregating distributed power capacity, energy storage, and other sources to optimize the power markets. The government intends in the future to promote VPPs in the country. VPPs will be allowed to participate in contracts, spot power markets, etc.

By 2030, the national VPP regulation capacity may reach up to 50 GW.

2.6 Early International Promotion of Renewable Energies: The Beginnings

2.6.1 The International Solar Energy Society (ISES): From Phoenix, Arizona, in 1954 to Freiburg, Germany, Today

2.6.1.1 From AFASE to ISES

In 1954, the NGO AFASE (the Association for Applied Solar Energy) was created in Phoenix, Arizona. It was followed just a year later by meetings in Tucson and Phoenix with a thousand participants from 36 countries, scientists, engineers, government officials, and a World Symposium on Applied Solar Energy the same year.

After being called the SES in 1963, it eventually in 1970 chose Melbourne in Australia as its headquarters, organized an international conference there and adopted a year later the name of ISES.

Then moved in the Europeans, the Italian C. Corvi from ENEL, the German A. Götzberger, the director of ISES in Freiburg. So in 1995 the ISES headquarters were moved to Freiburg, where it still is today.

Until this day, ISES organizes its Solar World Congresses at various places throughout the world.

In 2011, the author was honored by ISES with the **Global Leadership Award in Advancing Solar Energy Policy.** Monica Oliphant from Australia, a past President of ISES, took the initiative to create this new prize in honor of Hermann Scheer, who had just passed away the previous year. Since then, the prize is awarded every year by ISES.

2.6.1.2 The UK ISES from 1974

Lord George Porter, then Director of the Royal Institution in London, was promoting the creation of UK ISES in 1974. Participated among others Mary Archer, John Page, and Bernard McNelis; I attended the creation in 1974 as well. So did I in 2024 at the 50th Anniversary meeting at the Royal Institution in London.

Figure 2.20 September 2024. London, The Royal Institution. Celebration of the 50th Anniversary of UK ISES. First row from the right: William Gillett, a former colleague of the author from the EU Commission, DG Energy, then Mary Archer, Bernard McNelis, and the author.

2.6.2 The Conferences from the UN, New York

2.6.2.1 The United Nations Conference on New Sources of Energy, Solar Energy, Wind Power, and Geothermal Energy, Rome, August 21–31, 1961

This first UN Conference on renewable energies, decided by the UN Council, was indeed the go-ahead for a global development of solar energy and wind power in modern times.

It came just 7 years since the first efficient silicon solar cell had been invented in the US—and one of the inventors G. L. Pearson attended in person.

Likewise the conference had the privilege to have U. Juul as one of the organizers. He had built with others the Gedser Wind

Turbine in Denmark, "the grandmother of modern wind power plants" that had just started successful operation.

Fortunately, the UN has left by today the **proceedings** of this outstanding conference in digital form, seven volumes, some 1500 pages, in English and French. They give a comprehensive insight into the birth of the modern energies just after WWII.

The conference was attended by 447 delegates from 74 countries. For each subject, a Chairman had been appointed by the UN Secretary General, and a rapporteur had prepared in advance a general report on the subject and later a "summation" of the outcome.

Preparatory meetings were held in 1960 in Madrid, Grenoble in France, and in Rome.

A fine World conference, well prepared.

The part on PV comes in Volume 4 of the Proceedings, Solar Energy I. The following is a short summary:

- **G. L. Pearson,** one of the silicon solar cell inventors, now working at Stanford University, reported that the silicon solar cell they invented in 1954 was first used in October 1955 at Americus in Giorgia for a rural telephone; the PV module of 10 W and 22 V had 432 small cells.

  For its high cost at the time, PV had only a market for communication purposes, on Earth for wire or radio installations.

  PV was used since 1958, too, on all satellites, the first was Vanguard I, the second Sputnik III. For them, the PV cost was $1000/W.

- **M. B. Prince,** a co-worker of the group at Bell that invented the silicon solar cell, now working at Hoffman, one of the first solar companies, gave an overview of the latest PV developments. They had now succeeded in making the solar cells larger, 10 cm by 2 cm, and achieve 12% efficiency by applying metallic grids on the crystal surface. Cost has been decreased now in 1960 to $175/W from $500/W in 1957.

  He saw PV applications for community loudspeakers, highway emergency applications.

  They also worked on thin-film cells. With silicon, it was difficult. Most promising looked CdS. The company Harshaw achieved over 4% efficiency with it.

- **M. Wolf** from the new Heliotek Corporation, also saw wireless communication as the only market at the time. They worked on thin-film cells for the future, too.
- **P. Rappaport** from RCA Labs at Princeton, later the founder of NREL, had a look into the future. He thought for general applications on houses one had to look for PV at $1 per square foot or $0.02/kWh. A long-term issue; they also tried thin-film cells. And much work is needed, he thought, on storage.
- **M Kobayashi** from Nippon Electric Company reported that Japan started with PV in 1955. They saw a market for PV at $130/W when bringing a power line is more expensive. Since 1958 Nippon equipped many VHF repeater stations and Light Houses with PV: some 70 W each associated with NiCd batteries.

Volume 7 of the proceedings deals with wind power, in particular, "The design and testing of wind power plants." The session had 14 contributions. Its General Rapporteur was **L. Vadot** of the Study group in Grenoble, France, dealing with hydraulic applications. He presented a very competent report to start with. He reported to have in the session the presentation of six wind machines from 100 kW up to 1 MW. There were no fundamental problems with such turbines at the time of the conference.

The rapporteur noticed, however, difficulties with two-blade machines due to vibrations during orientation. Such difficulties disappear on three-blade rotors.

- **S. Armbrust** from the German Wind Power Research Association reported on the 100 kW machine designed by Hütter. A two-blader has been running at Stötten since 1955/57.
- **J. Juul,** gave a report on wind power plants in Denmark, in particular the building of the Gedser turbine.

In the session "Recent Developments and Potential Improvements in Wind Power Utilisation," **U. Hütter** was the Rapporteur. In his summation of the discussions, he noticed that the industrialized countries are now expected to undertake the considerable investments necessary for mass production of units.

2.6.2.2 Nairobi, August 10–21, 1981: UN Conference on New and Renewable Sources of Energy

This conference was oriented to the needs and opportunities of the Developing Countries. It came after a Resolution of the UN General Assembly in 1979, to provide assistance to the less developed countries.

It was organized by the UN Secretary General, with a Preparatory Committee, and Regional Commissions. Over a hundred Governments were involved in its preparation.

The Nairobi conference was attended by over 4000 delegates from 125 countries.

A Program of Action was adopted, stressing, in particular, the urgency to adopt renewable energies to solve the energy problems in rural areas.

2.6.3 The UN Environment Program (UNEP), Nairobi

UNEP recognizes the crucial role of fossil energy in climate change; for over 50 years, it has contributed to accelerating the transition to sustainable and affordable energy for all.

In its 2025 report, it noted for 2024 a blitz of important environment negotiations, such as

- the 6th UN Environment Assembly in Kenya
- the UN Climate Conference in Azerbaijan COP29
- the 5th session on legally binding instrument on plastic pollution in South Korea
- the UN Desertification COP16 in Saudi Arabia
- the UN Biodiversity Conference COP16 in Colombia

This Colombia conference gave no result. It was followed by another UN Conference on the same subject in Rome the following year. On its last day on February 27, 2025, at the last minute after harsh negotiation, an agreement was reached among the thousand participants from 150 nations—excluding the USA. COP16 adopted a 5-year work plan on saving the world's nature and biodiversity. A multi-billion-dollar budget was adopted to this end.

In **Gaza**, UNEP has found that the conflict has caused unprecedented levels of pollution, with sewage, debris, and toxic munitions contaminating soil, water, and the air.

UNEP also provides essential data on methane emissions with its international **Methane Emissions Observatory,** a satellite data and machine-learning system.

2.6.4 The Development Program UNDP, UN, New York

Through its Energy Moonshot, UNDP aims to catalyze and support the provision of access to sustainable, affordable and reliable energy to 500 million more people, which leads to green economies.

They announced to have an energy portfolio of 384 initiatives across 128 countries to provide some 80 million people with clean energy.

2.6.5 The United Nations Educational, Scientific and Cultural Organization (UNESCO) in Paris

2.6.5.1 UNESCO Conference on Improving Rural Living Conditions in the World, 1950

In 1950, UNESCO organized an early conference on improving rural living conditions in the world today via press, film and radio. Silicon solar modules had not yet been invented at the time. The Dutch had proposed then a sewing machine rotor with a bicycle dynamo and a NiCd battery for some light and radio reception for African countries. **The world was waiting for solar electricity!**

2.6.5.2 UNESCO Symposium on Wind and Solar Energy, New Delhi, India, 1954

In 1954, UNESCO organized a symposium on wind and solar energy at New Delhi, India. The conference was held in three languages: English, French and Spanish.

UNESCO published 2 years later in 1956 in Paris the proceedings of this major conference on the renewables: 238 pages, in English, today available in digitized form.

The solar part of the conference was mostly concerned with solar heat applications; PV was not mentioned much at that time, as the invention of the silicon solar cell, in the same year 1954, had just been published.

On the other hand, there was an extensive session on wind power. The world's global VIPs of wind power, J. JUUL from Denmark and U. Hütter from Germany attended. Machines up to 3 MW were discussed: This conference of 1954 was ahead of its time.

The conclusion was that many sites existed where large-scale deployment could be made and the output could be fed into the grid systems. "There is a possibility of production of electrical energy from wind power able to compete with that from steam plants": *This extraordinary acknowledgement of wind power's potential came from the pioneers as early as 1954!*

On October 26, 1954, **Dr. V. Baum**, from the Energy Institute of the Academy of Science in Moscow, a highly praised expert on the new energies, delivered the closing address at the symposium. He mentioned in particular: "Experiments and trustworthy calculations provide evidence that the harnessing of these new sources of power is worthwhile and commercially viable. The development of practical units for the conversion of solar energy and wind power into other forms of energy and the formulation of their introduction is imperative."

And further he said: "We can and must take advantage of the gifts the Sun showers on us. The task is a great and noble challenge to mankind."

This was seconded by Felix Trombe from Paris, who added: "After having destroyed part of his forests, man has found black coal and oil. He now thinks that some of these reserves of energy are running out as well, and that nuclear energy looks as a possibility of replacement. Its reign will not be long lasting either. And many of us think that renewable energies will accompany already nuclear development."

And ultimately our good old Sun and its energies will be the last to be with us.

2.6.5.3 UNESCO Solar Energy Conference, July 1973, Paris

In the first week of July 1973, another solar energy conference was held at UNESCO in Paris: "The Sun in the Service of Mankind."

The UNESCO headquarters in the center of Paris with nice facilities for meetings and a nice green environment lends itself well for the organization of a major event like that one we organized there in 1973. Over a thousand delegates from all over the world attended it.

Unlike in its previous conferences mentioned above, this time there was no session on wind power. The main French organizers of it, AFEDES and COMPLES focused on the Sun's thermal applications. Even solar PV was added to the program in extremis through a wakeup call from my boss Pierre Vasseur and myself from the French Space Agency CNES. Eventually, I became myself the organizer of its PV Session, and I organized and issued the proceedings of the PV Session that are still available today. At the time of that conference in 1973, there was a minor market only for PV, maybe a few MW a year besides those for the satellite application. But we all were full of optimism.

2.6.5.4 July 2013: A UNESCO Conference on the 40th Anniversary of the first one, "The Sun in the Service of Mankind" in 1973

We will come back later to the former 20th Anniversary Conference held in Paris in 1993 with its repercussions.

This conference, held in 2013 at the UNESCO House in Paris, was organized under the leadership of the French energy Agency ADEME. Dominique Campana from ADEME also organized the proceedings of the conference, which are available in French and in English in digital form, 120 pages each. The proceedings were a contribution to the UN initiative "Sustainable Energy for All (SE4ALL)" by 2030. The aim of that one is to ensure universal access to modern energy services, double the global rate of improvement in energy efficiency and double the share of renewable energy in the global energy mix.

Figure 2.21 40th Anniversary of the 1973 "The Sun in the Service of Mankind" Congress, held in 2013.

At this conference in 2013, I gave an opening speech stressing the recent achievements of the renewables by the time of the conference. *500 GW of PV, wind power and bio-power were installed up to 2013, 10 times of what it was at the turn of the century. In Europe under the lead of Germany its rate was 213 GW.*

320 GW of wind power, and 130 GW of PV, 100 times more of the latter than in the year 2000 were newly in operation worldwide; 100 billion liters of bio-fuels were produced globally in 2010 compared to 16 billion in 2000.

Just in 2013, $140 billion were globally invested in new wind and solar power development.

5.7 million new jobs had been created by then, 1.2 million in Europe.

For the year 2030 I anticipated some 2000 GW for wind power + PV (though optimistic, that was already achieved in reality in 2023).

By 2050, I declared a global 100% penetration of the renewables desirable and possible.

Leaders from REN21, IEA, and IRENA made projections for the renewables development by 2030 and 2050.

For China, F. Haugwitz from the Consultancy AECEA reported that the National Development and Reform Commission had just announced for 2017 150 GW for wind power, 330 GW of hydro, 70 million tons of biomass, and 70 GW of PV, half of it commercial and half distributed. This confirmed China's acceleration of the renewables' deployment in the country.

2.6.5.5 20th Anniversary Conference of "The Sun in the Service of Mankind," UNESCO, Paris, July 1993

The World Solar Summit Process: The above Conference had suggested undertaking a 3-year campaign 1993–1996 entitled "World Solar Summit Process (WSSP)." Eventually, the WSSP was adopted following their recommendations at a decision of the 27th Session of the UNESCO General Conference. The secretariat of the WSSP was held at UNESCO by B. Berkovski, Director of its Engineering and Technology Division.

The environmental imperative had been expressed just previously at the UN Conference on Environment and Development (UNCED), the Earth Summit of Rio in 1992. It was considered essential to develop environmentally sound technologies and industrial practices. In Agenda 21, it was agreed that energy is essential to development and quality of

life. There is an increasing need to rely on environmentally sound energy systems, the new and renewable energies, it said.

Figure 2.22 December 2023 in Paris: Boris Berkovski, former Director at UNESCO (left), Osman Benchikh, also a former official of UNESCO (middle), and the author (right).

The 1993 UNESCO Conference "The Sun in the Service of Mankind" had prepared 66 in-depth written assessments of critical solar energy fields such as

- solar energy and health
- solar energy, a strategy in support of environment and development
- financing solar energy development

These reports were debated in 35 specialized round tables. In conclusion, a **"Resolution addressed to the Founding Sponsors of the World Solar Summit"** was adopted.

It was agreed inter alia

- **that an International Solar Treaty for the global utilization of renewable sources of energy and efficient**

use of energy should be prepared. Its purpose was to encourage steps to develop, deploy and share the best renewable energy research and development data.

- **a World Solar Charter for**
  - the adoption of suitable educational criteria and enhancement of training;
  - preparation of awareness programs for the public concerning social, legislative, budgetary and technical incentives;
  - creation of a new form of ethics, on policy making, industrial planning, and standards of living;
  - promotion of investment and securing of financing for renewable energies, their global development and applications.
- **a World Plan of Action** for high-priority renewable energy projects to be implemented in the near future, such as
  - **Solar Energy for Rural Development.** The availability of modest quantities of energy, particularly electricity could substantially contribute to meeting essential health needs and improving quality of life by providing light, TV, telecommunications, water pumping and purification, refrigerators for vaccines—and to generate income through small industries. PV and wind power in particular may provide the minimum necessary power. They will have positive results in slowing urbanization, raising the level of education, and reducing demographic pressure. Actions have to be taken against the excessive traditional reliance on fuel wood.
  - **Public Information and Education.** There is great importance for the provision of **objective** information to the general public, as well as education and training on all levels.
  - **Solar Energy for Peace,** aimed at providing fresh water. The lack of water, irrigation, and fertilizers leading to insufficient food production—which were feared by the

end of the decade—are a major source of international tensions, destabilization, and conflicts. If implemented in cooperation with Middle East countries it could be a powerful incentive for peace.

- **Strategic projects,** in the field of energy policy, legislation, research and technology—to be performed in global collaboration. They were oriented towards helping the developing countries to overcome disadvantages they suffer in the field of energy supply and environmental protection.

Consequently, the following regional expert meetings were held:

- **Workshop "Solar Energy in the Mediterranean Region" UNESCO Headquarters Paris, July 12–13, 1993.** Participated delegates from Algeria, Egypt, France, Greece, Italy, Lebanon, Libya, Malta, Morocco, Spain, Tunisia, and Turkey. The meeting suggested the creation of a Mediterranean Network of Research and Training Institutions, MEDSOLAR.
- **Workshop "Rural Electrification in the Mediterranean Region" Marrakech, October 25–29, 1993, at the CDER Renewables Development Center.** The creation of a Mediterranean Solar Council and its inputs to the World Solar Summit were suggested.
- **An Action Plan for Renewable Sources in Europe, Madrid, Spain, March 16–18, 1994.** On the initiative of the Foundation Cànovas del Castillo and the EU, an Action Plan for the Development of renewable energies and their widespread utilization was prepared. It demanded the substitution of 15% of the conventional primary energy supply by renewable energy sources by 2010. This would lead to the creation of up to 400,000 new jobs, increase its turnover to over €6 billion and contribute to 350 million tons of GHG emission abatement.
- **Symposium "New, Renewable and Solar Energies in Africa," Bamako, Mali, May 16–20, 1994.** It was organized by the Organisation of African Unity (OAU) with delegates from Ivory Coast, Ethiopia, Ghana, Kenya, Mali, Niger, Sudan,

Swaziland and Zimbabwe. The meetings prepared inputs to the Harare Summit in 1996.

- **"Congreso LatinoAmericano sobre Energias Alternativas," Cordoba, Argentina, September 27–29, 1994.**
- **"7th International Symposium on Solar Thermal Concentrating Technologies" Moscow, Russia, September 26–30, 1994.** It was organized inter alia by the Scientific Committee for New and Renewable Energy Sources of the Ministry of Science and Technical policy of Russia.

To draw a conclusion, a major Governmental Congress was organized at Harare, Zimbabwe, September 16–17, 1996.

I attended it in person. Likewise, from the EU side, we supported the WSSP from the start.

2.6.5.6 50th Anniversary of the 1973 UNESCO Conference, July 2023

After 50 years, no wonder not many have survived from the original delegates of this big Congress.

I was just one of a few.

In particular, one had to regret the absence of any further interest in renewable energies, their developments and their enormous achievements from the UNESCO and its officials.

Figure 2.23 Dinner party in Paris in July 2023 on the 50th Anniversary of the UNESCO Solar Congress in 1973.

So just to take note of such a unique event, the 50th Anniversary of something colossal like solar energy, I took myself the initiative, supported by Daniel Lincot from the French CNRS, to invite some 50 experts to a memory evening at a nice place in Paris—and to review again all the achievements of solar development. The success did not come on its own.

2.6.6 The EU Promoting Renewable Energies: The First Steps

2.6.6.1 Cost-shared R&D program since 1975

In 1974, only a year after the UNESCO Conference of 1973 in Paris, the EU Commission—after an initiative of France—put together a first Solar Energy R&D Programme. To this end, the EU member countries—at the time the UK was a member but not yet Eastern Europe as the Soviet Union still existed—sent official delegations to outline a comprehensive Solar Energy Programme.

I came as a member of the French delegation. Since 1977, after the program had been adopted by a decision of the EU ministers in charge, I also became an official of the EU, its manager including the following ones up to 1998–99.

The first Solar Programme (1975–79) had a budget of €50 million. It was part of a wider R&D program including Energy Conservation Technologies, Green Hydrogen, and Energy Analyses, called Non-Nuclear Energy Programme or "Joule." Joule being part of the EU's 5-years Framework Programmes FP for Research.

The first European Solar Programme was divided into five parts:

- Project A, Solar Heating and Cooling
- Project B, Concentrating Solar Power Plants (CSP); the world's first CSP Eurhelios of 1 MW was built at Adrano in Sicily under my responsibility
- Project C, Photovoltaics
- Project D, Photobiology, Photochemistry, and Photoelectrochemistry
- Project E, Biomass

Calls for proposals were launched in all sectors; contractors had to be jointly from different member countries, part of funding was offered by the EU Commission, and the rest had to come from the contractors' own resources, or the budget of the contractors' country budgets, the EU member countries' ones.

Projects were further subdivided, for instance, such as "silicon cells" or "CdTe cells" for PV. There were numerous contractors from all over Europe, all leaders in their respective fields.

We were only a few at the EU offices in Brussels to manage the program, half a dozen co-workers with me, assisted by a contract department looking after contract administration and finance.

For program management, I adopted the "Contractors Co-ordination Meetings," held in Brussels and various places in Europe. Each contractor had to attend with a delegation. There were many of them, but we were able to handle it.

I was keen on looking "over the fence" and developed contacts with counterparts in the United States, Japan, and others. China was not yet very open in the early days of the 1970. To favor international interaction of our contractors, I initiated the first European PV Conference in 1977, **PVSEC**, the first Biomass Conference from 1980, **EUBCE**, and a European Wind Conference with EWEA in the early 1980. They all continue until this day, getting together thousands of experts every year.

The second 4-year Joule Programme started in 1979. We had a larger budget of €105 million this time.

Having learnt the lessons from the first program, we discontinued Project B on CSP, lowered efforts on Project D as results had not been promising, increased funding on Project C on PV, and started a new one on wind power.

A highlight of this program were the EU PV Pilot Projects built all over Europe for different applications, such as a PV plant on the NICE AIRPORT in France; they were all completed by 1982 and came long before the PV House markets were started in the early 1990s in Germany and Japan.

The following programs went on from 1985 to 1988 with a budget of €175 million, followed by one from 1989 to 1992 with €120 million. A particular highlight in these programs was the development of the first commercial MW wind turbines in

Europe via WEGA I and WEGA II, described in detail earlier in this chapter.

From 1993, we embarked on a special Solar Energy Programme called **APAS**. A budget of €25 million was availed to us by the European Parliament. Hundreds of stakeholders took part in a common effort to get all renewable energies closer to implementation.

This massive program was divided into five sectors:

- *Urban planning*: Towards zero-emission urban development
- *Decentralized PV electricity generation*: Study of large-scale manufacturing of PV. Cooperation with developing countries
- *EURENET*: European Regions' Network for Renewable Energies
- *Financial resources*: Assessment of international financial resources and opportunities for the development and use of renewable energies
- *PRODESAL*: Towards the large-scale development of decentralized water desalination

In 1993, I set up the **European Solar Council as part of APAS.** It became "The Club de Paris on Renewable Energy" as its first meeting was held there—at the Louvre actually.

The Solar Council linked the following EU networks that we had set up in the frame of APAS:

- EUREC, the EU Research Centres Agency on RE
- READ, a network of architects and building engineers
- CERE, the association of municipalities and towns
- EURE, the network of electric utilities.

In parallel to the APAS came another 4-year Joule Programme for the period 1994–1998. I had prepared it, and it was decided by the Council of EU Ministers on Research. **It demanded in particular "Actions by means of networks, several of which will be linked by a major network on the Development of RE."** It included inter alia networks on European electric utilities; pilot towns, regions, and islands; leading architects and building engineers; specialized solar research centers.

Figure 2.24 The solar village Pichling near Linz in Austria built under the auspices of the READ.

APAS and the Joule Programme from 1994 unfortunately could not be implemented as planned, as a new and incompetent official took over in Brussels under a public outcry and the dismantling of the Renewable Energy Division altogether.

A comparative study of non-nuclear energy research in Europe

This study commissioned by the EU Commission was implemented between 2003 and 2004. It gives an interesting insight into the particular role of EU energy research in the wider European Research Area ERA. It consisted of 33 country studies and three workshops.

It concluded inter alia that the only multilateral cooperation schemes for energy research were either provided by the European Commission through the RTD Framework Programmes such as the Joule Programmes, or by the IEA, through the Implementing Agreements. But unlike the EU Commission, the IEA doesn't dispose of its own budget to issue contracts, only to organize meetings in Paris.

At the time at the turn of the century, the European ERA countries together spent around €1 billion per annum on NNE-RTD. The EU Commission's budget—the Framework Programme for NNE-RTD—was about 1/5 of this sum. Japan spent about the same as Europe, whereas the US expenditures on NNE-RTD were at least twice as high.

Even though a great thematic variety existed among the European countries, there were shared priorities such as **fuel cells and solar PV.** To a lesser extent, there was interest in biomass and conservation.

Through its successive NNE-RTD Programmes, the European Framework Programme has been a major driver of multilateral research cooperation, this study concluded.

2.6.6.2 1997: EU White Paper "Energy for the Future"

I had authored it myself.

Communication from the Commission
"Energy for the Future: Renewable Sources of Energy"
White Paper for a Community Strategy and Action Plan
COM(97)599 final (26/11/1997)

Five years after the Rio Conference, Climate Change is again at the centre of international debate in view of the upcoming "Third Conference of the Parties to the UN Framework Convention on Climate Change" to be held in Kyoto in December 1997. The EU has recognized the urgent need to tackle the climate change issue. It has adopted a negotiation position of 15% greenhouse gas emissions reduction target for industrialized countries by the year 2010. To facilitate the Member Countries achieving this objective, the Commission, in its communication on the Energy Dimension of Climate Change from May 1997 identified a series of energy actions—including a prominent role for the Renewables. **The Council of Ministers endorsed this when inviting the Commission to prepare an action plan and present a strategy for renewable energy.**

Current trends show that considerable technological progress of the renewable energies has been achieved in recent years. Costs are rapidly dropping and many Renewables have reached or are approaching economic viability. The first signs of large-scale implementation are appearing as regards Wind Energy and Solar Thermal collectors. Biomass, Small Hydro and Wind are currently competitive and economically viable in decentralized applications. Solar PV, although with declining costs remain more dependent on favorable conditions. Solar water heaters are currently competitive in many parts of the Union.

Energy prices for conventional fuel cycles do currently not reflect the objective full cost, including the external cost to society of environmental damage caused by their use.

For the new Renewable Energies global turnover of the industry is estimated to be higher than €5 billion, a third of it for Europe.

Three policy objectives were identified as regards a Community energy policy: improved competitiveness, security of supply, and protection of the environment.

As the internal market develops, a Community-wide strategy for Renewables is required to avoid imbalances between Member States or distortion of energy markets.

The leading position of the European Renewables energy industry can only be maintained and strengthened on the basis of a significant and growing home market.

A policy for the promotion of Renewables requires across-the-board initiatives encompassing a wide range of policies. Energy, Environment, Employment, Taxation, Competition, Research, Technological development and demonstration, Agriculture, Regional and External Relations Policies.

A comprehensive action plan is required to ensure the necessary co-ordination and consistency in implementing these policies at Community, National, and Local levels.

Member States and the Community should formulate indicative targets as guideline for the ambitious target of doubling the overall share of the Renewables in Europe by 2010.

The European Parliament believes that the EU urgently needs a promotion strategy to tackle the issues of tax harmonization, environmental protection and standards, internalization of external costs, and ensure that the gradual liberalization of the internal market will not place the Renewables at a disadvantage. It proposes a goal of 15% share of the Renewables for the EU by 2010.

The strategy and action plan in this White Paper are directed towards the goal of achieving a 12% penetration of the Renewables in the EU by 2010.

The current share in 1997 of the Renewables in the energy mix of approximately 6% includes large-scale Hydro.

The scenario as outlined here below, assumes

- **The main contribution of RES growth (90 million tons) could come from Biomass, tripling the current use of this source. 15 million tons could be gained from new biogas exploitation (livestock production, sewage treatment, landfills), 30 million tons from Agricultural and Forest residues, and 45 million tons from Energy crops** Currently Biomass accounts for about 3% of total inland energy consumption in the EU15. The 3 new Member States Austria, Finland and Sweden have a much higher share of it today. Additional Biomass of 90 **million tons** is equivalent to 8.5% of the projected total energy consumption by 2010.

- **Wind energy, with a contribution of 40 GW, is likely to have the second most important increase.** Europe is the world leader in Wind Power with more capacity installed than any other region in the world: 3.5 GW in EU 15 at the end of 1996. If current installation rate continues the following years 18 GW will be installed by 2010, but if production increases every year at the same rate like in the past we'll have 100 GW of Wind Power already. **An estimated increase to 40 GW is a conservative figure.**

- **Significant increases in the Thermal collector market (100 million m^2 installed by 2010) are anticipated.** The current annual installation rate is 1,000,000 m^2. An annual growth rate of 20% is anticipated to reach 1 million m^2 by 2010,
- **PV could increase to 3 GW.** Costs have fallen 25% over the past 5 years. The EU accounts in 1997 for a third, 100 MW, of global PV module production and use. World annual production is forecast at 2.4 GW by 2010. Of the 3 GW installed in 2010 in EU 15, most of it would be grid-connected, either rooftops or as large-scale power plants. A cost level of €3/W is anticipated.

A campaign for take-off promoting the installation of 1.000.000 PV roofs and façades would mean a new capacity of 0.5 GW installed in the EU and 0.5 GW in third countries

- **Finally Passive Solar could make a major contribution in reducing heating and cooling demand in buildings. A 10% contribution in this sector, representing fuel savings of 35 million tons is considered feasible.** Thermal energy demand mostly for space heating In the domestic and tertiary sector in EU 15 represents 23% of total energy demand. It is estimated that 40% of energy actually consumed in this sector is in fact gained from Solar energy through windows, but this passive energy supply is not taken into account in statistics. Experience in Austria has shown that passive Solar construction increases overall dwelling costs by less than 4% while achieving 75% reduction in heating energy. A 10% reduction by 2010 of thermal energy in buildings is easily attainable.
- **Heat pumps, 2.5 GW$_{th}$ by 2010.** A total of 60.000 heat pumps were installed in the EU in 1995, most of them in Sweden, representing 8% of the capacity of all types. We

assume a tripling of the total installed heat pump capacity by 2010, and the share of Geothermal heat pumps to double to reach 15%

As far as the market sectors are concerned, a doubling of the current electricity and heat production from the Renewables plus a significant increase of Bio-fuels in transport fuel use by 2010 are important elements of the scenario.

Electricity from the Renewables could grow significantly from the present 14.3% to 23.5%. Heat production from the Renewables is supposed to double.

2.6.6.3 What happened after the EU White Paper from 1997: How good were the projections

What was indeed achieved by 2013 when an account was published on the occasion of the 40th anniversary of the UNESCO Conference of 1973 in Paris? See also the report there.

In 2013, new 510 GW of renewable energy was installed (solar PV, wind power, and bio-power) compared to 47 GW in the year 2000. In the EU, 213 GW was achieved with Germany, 75 GW, in the lead.

Year-end 2013, 130 GW of PV was globally installed against 1.4 GW in 2000, a factor of 100. Wind power in the world came an impressive 320 GW.

Bio-fuel markets (bio-ethanol and bio-diesel) went from 16 billion liters in 2000 to 100 billion liters in 2010.

5.7 million new jobs have globally been created, 1.5 million of them in Europe and 380 thousand of them in Germany.

2.6.7 The 1979 EU Conference, "The Sun in the Service of Development"

In March 1979, on behalf of the European Commission, I organized the Governmental Conference "The Sun in the Service of Development" in Varese, Italy.

The conference was attended by official governmental delegations from 80 nations from all over the world, several UN organizations, international banks and institutes. It was opened by Sir George Porter, a Nobel Prize laureate and director of the Royal Institution in London.

Various regional seminars led up to the full conference. I was active in preparing and chairing them and even wrote out the bank cheques for the invited participants on behalf of the EU Commission. The seminars were the following:

- September 1978, **Nairobi, Kenya, for East Africa,** with official representatives from Kenya, Ethiopia, Lesotho, Mauritius, Rwanda, and Zaïre (Congo). It was opened by the Kenyan minister for water development.
- September 1978, **Bamako, Mali, for West Africa,** with representatives from Mali, Mauretania, Niger, Chad, and Togo. Three ministers from Mali and Upper Volta (Burkina Faso) participated.
- October 1978, **Amman, Jordan, for the Arab World,** with delegations from Jordan, Syria, Sudan, Saudi Arabia, Kuwait, and the Arab Physical Society. It was also attended by the Crown Prince of Jordan and 3 of his ministers, 3 institution presidents, 10 director generals and 6 ambassadors.
- October 1978, **Caracas, Venezuela, for Latin America** with delegations from Venezuela, Argentina, Bolivia, Brazil, Chile, Mexico, Peru, the Latin American Solar Energy Society, and OLADE, the Latin American Energy Organization.
- October 1978, **New Delhi, India, for Asia** with representatives from India, Pakistan, Sri Lanka, Singapore, Bangladesh, Indonesia, and Thailand. It was opened by the Secretary for Science and Technology of the Indian Government.

The full Conference in Varese stressed the need for international cooperation in research, transfer of technology, education and capacity building, as well as environmental and social aspects. As far as PV was concerned, it was interesting to note that some participants expressed a reluctance to depend on imports from industrial countries for their supply. **The typical argument: This is our Sun!**

2.6.8 The First World International Renewable Energy Conferences (WIREC)

2.6.8.1 Bonn Renewables 2004

In the frame of Agenda 21 adopted in Rio in 1992, and following the Earth Summit in Johannesburg, South Africa in 2002, the German Government convened to the first WIREC, a political conference entirely dedicated to solar energies.

It had some 3000 delegates from 154 countries. H. Scheer was its main operator. My friend Michael Eckhart from Washington was together with me one of the drivers for its success.

The conference concluded rightly that renewable energy will be the most important and widely available future energy system. Needless to say, renewable energy will not only be the most important, but it will be the most important energy system—and not only in future, it is here today.

2.6.8.2 Beijing 2005

The WIREC at Beijing was held at the Great Hall of the People. M. Eckhart President of the American Council on Renewable Energy (ACORE) attended and so did I myself. As mentioned already elsewhere, 2005 was also the year in which the People's Congress of China adopted a promotional law on the renewables with a FIT for PV.

2.6.8.3 Washington 2008

Thanks to M. Eckhart, the WIREC in Washington, DC, became a great success, too. President Bush came with several of his cabinet ministers and gave a speech. It was a great mobilization of the US interest in the renewables. Eighty pledges were adopted to do more for its development in the future.

The conference ended in a snowstorm, but that is another story...

2.7 Another "Clean" Emission-Free Power: Nuclear Energy, a Revival?

At the COP 28 Climate change Conference in December 2023 more than 20 countries signed a joint declaration to triple nuclear power by 2050. That would globally mean an addition of 740 GW of nuclear capacity to the current stock of 370 GW.

My comment: not in my name!

2.7.1 Modern Nuclear Power Plants

Flamanville 3 in Normandy, France, an EPR-1650 of 1.65 GW net capacity: The last built atomic power plant in France was connected to the grid in late 2024. The start of commercial operation was planned for January 2025: but in April it was still not grid connected owing to technical problems; in the summer, maybe? No, neither in summer. Maybe after autumn?

Its cost had been scheduled at €3.3 billion for a construction time of 5.1 years. But at the end, construction time was 17 years, a 12-year delay, and total cost was €19.9 billion—and not €13.3 billion that one excluding financing and cost overruns. The French Court of Audit concluded in 2020 already on €19.1 billion of total cost. Eventually, in 2025, the total cost of over €23 billion was found.

The cost per Watt came out to be €17.3, all government paid, taxpayers' money in a country that is one of the most indebted of Europe. At a time when PV module cost came some €0.10/Watt.

And there is the older Flamanville 1, which remained out of order until May 2025 owing to a nuclear accident, the outflow of radioactive water.

No comments.

Hinkley Point C 1-2 in Somerset, England, two EPR units of together 3.2 GW: The plants are built by the French EDF and 70% financed by France.

The plants are scheduled to get operational in 2031. In an estimate of 2024, they would have cost €46 billion, €14.175/kW; originally it was scheduled at €2.324/kW. The realized construction time has reached from 11.7 to 17.7 years so far; more delays are expected.

Radioactive waste disposal: Following the French Agency for radioactive waste management ANDRA—with new figures published in May 2025—the cost of France's Cigéo, the underground storage plant, was evaluated between €26.1 billion and €37.5 billion. As it would be fully operational only in 2050, French experts evaluate its final cost to be up to €200 billion.

By 2025, the world had already produced 263,000 tons of radioactive waste. A lot, and no final storage site in view...

In the old days, a lot of waste barrels were sunk in the oceans, where they are rotting until this day, and not only a few: Until 1993, 200,000 barrels had been sunk in the North Atlantic, endangering life conditions there.

In Europe and America, no other modern atomic power plants are being built in 2025, fortunately, as plans are far from being reliable as they purposely omit large overruns in cost and construction times. kW costs are over 10 times the cost of modern PV and wind power plants.

Time-wise, they come too late to help alleviate the pressing growth of the climate change effects that we have to fight, not only in 10 or 20 years but today. Deploying PV or wind power plants takes months, not years.

2.7.2 The Other Nuclear Power Plants

Let's have a quick look at the other atomic plants that are in operation today. Most of them have been built in the last century, a few are built today in Asia, most of them in China. But those in China play only a minor role; at year-end 2024, new atomic power capacity stood at 61 GW in China, 2% of the country's capacity, compared to 1,407 GW for PV and wind power.

We refer in the following to *The World Nuclear Industry Status Report 2024*, by Mycle Schneider in Paris (see also his separate article in this book):

In 2024, globally there were 411 reactors of 371 GW total capacity in operation. In 2023, they had produced 2.602

TWh of electricity, some 9.15% of the world's total electricity supply. There was agreement until recently—before governments took measures to extend operational years, increasing a risk of accidents—that some 40 to 60 years is the lifetime of these old nuclear plants. Currently, the mean age of all nuclear plants in France is already over 39 years; in the United States, the nuclear plants together are some 44 years old, and no new ones are being built. It would be a Herculean task to dismantle approximately 200 plants, leave alone the tremendous costs of dismantling, all this radioactively contaminated material, disposal of all the used fuel; it remains highly radioactive for centuries.

Germany, the EU's biggest economy, has no more any plants running. Following the accident in Japan, it decided to completely leave the nuclear plant business.

Between 2005 and 2024, in China there had been 51 startups of nuclear plants; outside China, the world had 53 startups and 101 closures. The EU 27 had 35 closures and 4 startups, those described here above.

The master of the "nuclear game" today is Russia. In 2024, it delivered 573 tons of enriched uranium to the EU, 38% of its needs. It dominates the export of new plants under construction in India, Egypt, Turkey, Iran, Slovakia, ...

2.7.3 The Small Modular Reactors (SMRs)

Commercial SMRs don't exist today, nor are they expected to make it into the global markets in the next 10 years or so. It is estimated that tens of thousands of them would have to be built to make an impact. Early startups in the United States gave up again when the **lack of competitiveness with respect to modern PV or wind power** was understood.

But as a matter of fact, many countries and their politicians around the world find them promising and invest in them. A lot of disappointment is waiting around the corner, with another mountain of taxpayers' money to be wasted.

Since the early days of nuclear power in the 1950s, with the "Army Nuclear Power Program" in the US for instance, small reactors looked interesting, but eventually the larger plants, by 5 to 10 times larger, up to 1.5 GW capacity, made it into the markets for an economy of scale.

So far, SMRs have been successful on ships, naval reactors for turbo-shaft props. Some hundreds of those have been built. The famous ones are the Russian nuclear Ice Breakers.

2.8 Sustainable Development, Now!

2.8.1 The UN Sustainable Development Goals (SDG): Stock Taking in 2025

In May 2025, the United Nations published a report on the progress made since 2015 towards its Sustainable Development Goals. It uses inputs from more than 50 international and regional organizations to the data framework of global indicators, altogether 137 of them. It accounts for the transformative change at the country level on

- Food systems
- Energy access and affordability
- Digital connectivity
- Education
- Jobs and social protection
- Climate change, biodiversity loss and pollution

From the many details, just a few essentials are summarized here:

- Since 2015, extreme poverty has declined, but the number of some 700 million people concerned is still very high!
- Hunger and food insecurity have escalated with 750 million people hungry and 2.3 billion facing food insecurity. Government spending on agriculture is insufficient.
- Global GDP per capita has rebounded since 2020.
- Internet usage reached 68% of the global population in 2024 compared to 40% in 2015; 2.6 billion people remain offline.
- There was progress on healthcare; childbirth, HIV infections, malaria and other diseases.

- Electricity access rose from 87% to 92%. Access to safe drinking water, basic hygiene and sanitation has improved. Access to clean cooking fuel has increased, too.
- Peace and security, the foundations of sustainable development, have declined sharply.
- Carbon dioxide levels stand at 151% above the preindustrial levels, the highest in 2 million years. Ocean acidification and species extinction risk continue to rise.
- Financial flows to developing countries have increased, from both official and private sources.

Progress on the SDGs falls far short of the scale or speed needed, the report concludes, and "the 2023 SDG Summit Political Declaration and the 2024 Pact for the Future reaffirm sustainable development as a core pillar of multilateral cooperation."

2.8.2 The Solar Energies for Climate and Peace, the Royal Route for a World Out of Distress: Comments from the Author

Nowadays, a large majority of people around the world, also in the industrialized countries, are considered poor; women and children are suffering the most from poverty. In the OECD countries, 2 million people are even homeless; in Los Angeles, like in Paris, a lot of people live in the streets.

The benevolent help as well as they can; without the food banks, for instance, in Germany or England, hunger would be widespread in Europe. The UN counts 2.4 billion people worldwide suffering from food insecurity.

Numerous conflicts and wars are raging in many countries, dozens of them, not only in Sudan, Ukraine or around Israel. As a rule, industrialized countries are not directly involved; they just provide the arms—a big business. Who suffers the most from the bombings? innocent women and children.

The effects of climate change make things worse, just to mention the droughts and wildfires, the weather disturbances

and the inundations. Today, insurance companies around the world already pay billions of dollars for all the damage.

This book has shown how solar PV, wind power and all the other renewable energies have already succeeded today on a large scale in alleviating GHG emissions and climate change.

Remember the question when a glass is half full of water—or is it half empty? **In energy terms, virtually half of the job is done: Half of the World's power today in 2025 is already renewable. By 2028, over 50% of the electricity consumed by the world will also be renewable.**

This way, income and wealth are created for all, along with jobs, trade, education, entertainment, tourism, and small industries. And there is enough of it for all—no reason for conflicts.

In countries like India or China, the pollution through the excessive use of coal may have regressed. Examples of developing nations where people benefitted at last from a reliable source of power, such as Pakistan and South Africa, are real.

2.9 Conclusions

The decade in which we all live is a special one. By now it is only 5 years old, but what a decade... It started in 2020 with a worldwide pandemic—COVID-19—lasting 2 years and bringing everything to a standstill.

For renewable energies, it was the time when they became grown up. And by now "young solar and wind" have pushed the "grandfathers coal and nuclear" to retirement. And it was time: The world started to suffocate from pollution it raises in never-ending quantities together with all those greenhouse gases.

Solar PV and wind power now in the middle of this decade play in the upper league. Their power capacities are now counted in TW, 1000 GW, or 1,000,000,000 kW, no more "the little GW" only. When our decade started, PV and wind power stood at some 1.3 TW, but only when counted together. Each of them alone made about half of it, PV at some 600 GW in early 2020 and global wind power at 700 GW or so.

Now, some 5 or 6 years later, PV capacity has achieved 2.8 TW worldwide and wind power 1.2 TW, together some 4 TW, a multiple of what it had been only a few years ago. And their exponential rise is about to continue, if not to accelerate.

And you like it or not—readers of this book take note with enthusiasm—all renewable energies, derived from our good old SUN, under the leadership of PV and wind power, dominate the power markets now. Their global installation rate together with their new partner battery storage, predominates new installation rates globally by over 90%. No matter the few new coal and nuclear plants in China, or a new president in the United States, who just recently in April 2025, signed an order to promote coal plants again in his country.

Renewable energies make these significant headways in the electricity sector, just where incremental supply is now needed. Over 5 billion people use smartphones today, for example; the number was 50% smaller at the beginning of our decade. So this is not more need for light, but rather new data centers that are driving more power demand.

Apart from bio-energy—for instance, if you are a car driver, you get most of the time some bio-alcohol with your gas—renewable fuel and heat will come as well into everybody's field of interest. Be it for hydrogen or ammonia derived from the renewable power or solar heating and cooling for working and living quarters. It is only a matter of time.

True, the efforts the world has still to make in the energy field, outside power and electricity, are certainly gigantic: all the fuel-burning vehicles in transport on the roads, on the seas and in the air. Cement, steel, and ammonia for fertilizers—see the early projects of the Middle East nations. The agenda is now well understood.

And where will the world go from here?

HELP! The world is experiencing climate change. Already in 2024, the mean global temperature was +1.5°C above preindustrial times, enormous! And all the climatic consequences—fires, inundations and all the rest—are increasing with great speed; +2°C is just around the corner: Climate change in the midst of our decade is in full swing. It is like in the pandemic a few years ago, everybody is concerned, some more, some less.

For all of us, the polluters of today, there is an ethical imperative to help alleviate it; in particular with respect to the coming generations.

To keep the GHG emissions at bay, the deployment of renewable energies has already well shown its capabilities; giga tons of CO_2 emissions were avoided—just in the last few years. **Further deployment of PV, wind power and the other energies derived from the SUN is the way forward.**

But the world is inhabited by all kinds of people, and the rational way of thinking is not in everybody's mind. Just now, America started some commercial conflicts, and with new political insecurity, Western Europe is embraced even by war hysteria. It is for sure that Donald Trump makes the world tremble, even his rich friends: In April 2025, Wall Street went down the drain as much as it had lost last time at the beginning of our decade in June 2020...

Eventually, in May 2025, Trump presented himself also as champion of nuclear power: executive orders to build more nuclear in his country, and faster. Please don't laugh.

Stock markets always recover, and so they did this time, too. Confidence in the US president was increasing, leave alone his incompetence and bad will for the renewables. And his lack of recognition of climate change, strange as it is.

It must be feared that a motivation to fight global change as such may not be good enough to keep today's enthusiasm for the solar energies. "If you don't like the SUN, go to hell" will not work, in particular for all those investors who still spend billions of dollars to dig coal, oil, gas and uranium out of the ground...

Cheers, there is still a common denominator for all: money. **Money makes the world go round!**

Were the clean renewables not competitive in the global markets—as was the case only some 10 years ago—you could almost forget them, how important they may be to save the world. But here you are: they are the cheapest. Their deployment results in not only cleaner air and cleaner water and stabilization of the climate but also better welfare of people with reliable power everywhere, more jobs and more money in pockets.

One thing is for sure: The solar energies will continue their breathtaking speed, for the good of all of us.

Chapter 3

China's Wind Power, Half a Century of Forging Ahead

Qin Haiyan

Secretary General, Chinese Wind Energy Association (CWEA), China

qinhy@cwea.org.cn

One day in September 1990, He Dexin, who was working at the China Wind Energy Technology Development Center, received a letter from Qian Xuesen, a famous Chinese scientist.

In the letter, Qian Xuesen outlined a plan for China's 21st-century power generation dominated by wind power. One specific goal is to build "4 million standardized 500 kW wind turbines." Qian Xuesen also gave specific suggestions on how to achieve this goal: the first is standardization to reduce costs and improve reliability; the second is the selection of sites and grids, starting from the use of existing facilities; the third is other issues, such as the organization of mass production of equipment.

Qian Xuesen concluded his letter with the following instructions: "Wind power cannot be a small business in the future. To do big things start by raising your own awareness! This will be the Socialist China in the 21st century."

The Modern Energies
Edited by Wolfgang Palz

ISBN 978-981-5352-02-3 (Hardcover), 978-1-003-79730-2 (eBook)
www.jennystanford.com
DOI: 10.1201/9781003797302-3

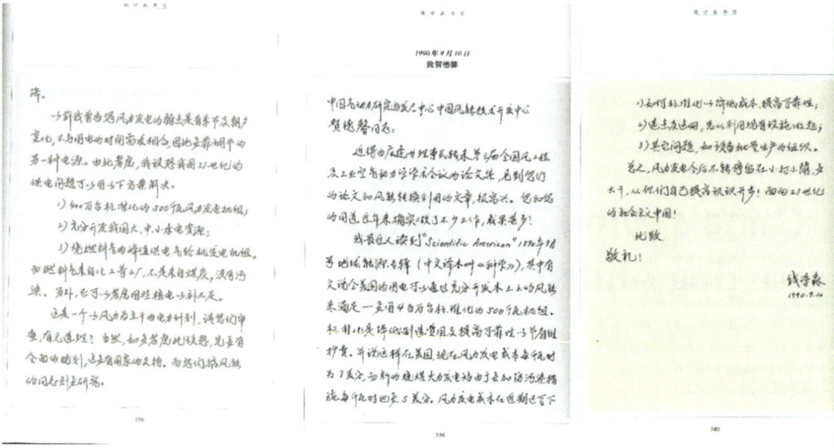

1990年9月10日
致贺德馨

中国空气动力研究与发展中心中国风能技术开发中心
贺德馨同志：
近得由广东汕头理事长转来第三届全国风工程及工业空气动力学学术会议的论文集，见到您们的论文和风能转换利用的文章，很高兴。您和您的同事近年来确实做了不少工作，成果甚多！
我最近又读到"Scientific American"1990年9月号地球能源专辑（中文译本叫《科学》），其中有文说全美国的用电可以通过充分开发本土上的风能来满足——要用4百万台标准化的500千瓦机组。标准化是降低制造费用及提高可靠性与节省维护费。并说这样在美国，现在风力发电成本每千瓦时为7美分，而新的烧煤火力发电站由于要加防污染措施每千瓦时也要5美分。风力发电成本在近期还可下

降。
以前我曾知道风力发电的缺点是有季节及每日变化，不与用电的时间需求相合，因此要靠调节的另一种电源。由此考虑，我认为我国21世纪的供电问题可以用以下方案解决：
1）400万台标准化的500千瓦风力发电机组；
2）充分开发我国大、中、小水电资源；
3）烧燃料气的峰值供电气轮机发电机组。而燃料气来自化工营业厂，不是来自煤炭，没有污染。另外，也可以考虑用些核电以补不足。
这是一个以风力为主干的电力计划，请您们审查，有无道理？当然，如要考虑此设想，先要有全面的规划，还要有国家的支持。而您们搞风能的同志则应研究：

1）如何标准化以降低成本、提高可靠性；
2）选点及送网，先以利用现有设施做起；
3）其它问题，如设备批量生产的组织。
总之，风力发电今后不能停留在小打小闹，要大干，从你们自己提高认识开步！面向21世纪的社会主义中国！
此致
敬礼！
钱学森
1990.9.10

Figure 3.1 1990, Qian Xuesen's letter to He Dexin.

In September 1990, at the "New energy (power generation) work Symposium" held by the Ministry of Energy, He Dexin conveyed Qian Xuesen's views, which caused a warm response from the conference representatives, and Huang Yicheng, then Minister of Energy, clearly pointed out at a wind farm construction planning meeting held by the Ministry of Energy: "We should lose no time to put wind power on the agenda, implement unified planning and organizational leadership, mobilize the enthusiasm of all aspects, and make wind power become a front army in China's power industry."

The letter is just one of 18 personal letters Qian Xuesen wrote to He Dexin between 1982 and 1996. These letters involve China's wind power development ideas, wind energy product discussion, technical verification and calculation, etc. He Dexin has been devoted to China's wind power industry for decades, promoted the basic theoretical research of wind energy and wind power product innovation, and established industry organizations such as China Wind Energy Technology Development Center and Chinese Wind Energy Association of China Renewable Energy Society, serving the goal and strength of national wind power development planning and strategic research.

Figure 3.2 He Dexin has been devoted to China's wind power industry for decades.

Since then, China's wind power installation capacity has soared higher and higher, running stronger and stronger on the road of "wind power powerhouse," becoming the wind vane of the world's wind power, and achieving remarkable achievements. By 2010, China's cumulative installation capacity of wind power reached 41.827 million kilowatts, surpassing the United States and ranking first in the world. In 2024, China added 86.99 GW of newly installed wind power capacity, bringing its cumulative installed capacity to 561.26 GW. Wind power now accounts for 15.2% of the country's total power generation mix. It is expected that by 2030, China's cumulative installation capacity of wind power will reach 1 billion kilowatts, and it is expected to reach the installation capacity target of 2 billion kilowatts in Qian Xuesen's letter by 2040, and gradually become the main energy source in China's new power system.

3.1 Start the Policy Engine

From regulations to policies, from norms to planning, China has built a comprehensive, accurate and effective wind power management system. It is through the continuous guidance of this management system, coupled with the efforts of all wind energy people, that the scale of wind power development in China has increased by more than 1,200 times.

If the time goes back to 2002, when China's cumulative installation capacity of wind power was only 420,000 kW, wind turbines were still mainly introduced from abroad, and the single capacity was at the 100 kW level, and the entire Chinese wind power industry had not embarked on the road of large-scale development. It was in that year that Shi Lishan left the hydropower industry and began to work on renewable energy management, until he was transferred to the deputy director of the Nuclear Power Department of the National Energy Administration in 2016, a total of 13 years in this position. In the past 13 years, it has become the fastest growing stage of China's wind power industry.

Figure 3.3 Shi Lishan delivering a speech at a wind power-related conference.

The wind power industry recognizes Shi Lishan as the chief architect of China's wind power policy. From his tenure as Director of the New Energy Division at the Energy Bureau under the National Development and Reform Commission (NDRC), to serving as Deputy Director-General of the Department of New and Renewable Energy at the National Energy Administration (NEA), Shi Lishan dedicated nearly 2 decades of his public service career to advancing renewable energy development—including wind power—through close collaboration with colleagues on policy design and implementation The key policies of each critical period of the wind power industry are almost all introduced under the organization, research and revision of Shi Lishan, and no one knows more than him about China's wind power policy and the development trajectory of the wind power industry.

3.1.1 The Cornerstone of Legalization

"From small to large, two initiatives have played an important role in China's wind power industry, one is the 'Renewable Energy Law of the People's Republic of China,' the other is the wind power concession bidding, the former is the legalization of China's wind power development cornerstone, the latter is to promote the large-scale development of the wind power industry policy engine."

Looking back at the development of wind power in China, the wind power industry is also like life, as the Chinese writer Liu Qing said in the novel *History of Entrepreneurship*, "Although the road of life is long, but the important thing is often only a few steps, especially when people are young."

Shi Lishan believes that from small to large, two initiatives have played an important role in China's wind power industry, one is the "Renewable Energy Law of the People's Republic of China," the other is the wind power concession bidding, the former is the legalization of China's wind power development cornerstone, the latter is to promote the large-scale development of the wind power industry policy engine.

The Renewable Energy Law of the People's Republic of China, with its short legislative time, timely legislation, effective legislative system, special legislative process, are incomparable to other legislation, efficiency "in the history of legislation is very rare."

The push for the Renewable Energy Law of the People's Republic of China, which Shi Lishan likens to "a very good gift from God," has almost coincided with the bidding for wind power concessions.

This law was legislated by the Environmental Protection and Resources Conservation Committee (EPRCC) of the National People's Congress (NPC), speeding up the pace of legislation, began drafting in 2003, was adopted at the fourth session of the Standing Committee of the Tenth National People's Congress on February 28, 2005, and came into effect on January 1, 2006. Its short legislative time, timely legislation, effective legislative system, special legislative process, are incomparable to other legislation, efficiency "in the history of legislation is very rare."

Figure 3.4 The Renewable Energy Law of the People's Republic of China which came into effect on January 1, 2006.

In Shi Lishan's mind, the fundamental purpose of the "Renewable Energy Law of the People's Republic of China" is to encourage renewable energy and create a legal environment for the vigorous development of renewable energy, which refers to the development of the wind power industry to protect the legal level, feed-in tariff, the allocation of the whole grid and full acquisition is its core content. Among them, the feed-in tariff, and the full acquisition of the whole grid, for investment enterprises, as long as the construction of wind power projects, they can be sold to the power grid in accordance with the feed-in tariff, of the project, in order to calculate the income of their investment, solve the worries, and mobilize the enthusiasm of investment enterprises. The allocation of the whole grid has mobilized the enthusiasm of local governments to support wind power, because the part that is higher than the average electricity price is borne by the state, rather than the local government.

In order to keep the Renewable Energy Law of the People's Republic of China up to date, it was revised at the 12th session of the Standing Committee of the 11th National People's Congress on December 26, 2009. In 2023, the formulate (amendment) of the Renewable Energy Law of the People's Republic of China was started again. Commissioned by the National Energy Administration, the Energy Institute organized various units to carry out research activities on the "Revision of the Renewable Energy Law" project, and the Chinese Wind Energy Association organized enterprise research and carried out seminars to form a research report on the revision of the Renewable Energy Law and a special report on the coordinated development of land, sea and ecological environmental protection. At the same time, actively participated in the field investigation in Guangdong and Yunnan Province and formed corresponding research reports.

In addition to the Renewable Energy Law of the People's Republic of China, on November 8, 2024, the 12th session of the Standing Committee of the 14th National People's Congress voted to pass the Energy Law of the People's Republic of China, which will come into force on January 1, 2025. It is the basic and dominant law in China's energy field, focusing on the major policies, fundamental principles and institutional systems of China's energy work. It further establishes the development status of renewable energy, clarifies the state's support for the priority development and utilization of renewable energy, and accelerates the replacement of renewable energy by fossil energy through institutional arrangements such as the formulation of medium and long-term development goals for the development and utilization of non-fossil energy and the minimum proportion of renewable energy in energy consumption.

3.1.2 Special Policies for Special Times

In Shi Lishan's view, wind power concession bidding is a special policy in a special period. It is through the selection of conditional wind farms, the government promises to implement the relevant construction conditions, the power grid enterprises promise to implement grid connection and full acquisition of the generated electricity, through public bidding, with the on-grid tariffs and wind power equipment localization rate as conditions, select the wind farm construction investors. Its main purpose is to promote the large-scale construction of wind power, the localization of wind turbine manufacturing and the industrialization of wind power.

Before the launch of the wind power concession tender, the wind turbines used in China's wind power market were almost all imported products, and the market scale was small, and the average annual new installation capacity in 2001 and 2002 was only 60,000 kW, which was far from the good expectations. This is because the price of wind power is very high, the price of electricity approved by the price department is 1.5 yuan/kWh or even 2 yuan/ kWh, so expensive electricity no one is willing to use, and it will lose the opportunity for development. It is such a status quo and reality, Zhang Guobao, then deputy director of the National Development and Reform Commission and director of the National Energy Administration, put forward the proposal of "expanding the pie, promoting localization and introducing a competition mechanism." What Shi Lishan has to do is to put these three aspects of the proposal specific to the concession bidding policy, and leave room for optimization and improvement, so that it becomes China's wind power scale cost reduction engine development.

As a contract between the tenderer and the bidder, the concession project bidding has changed the previous routine of the government building wind farms. Whoever invests in the project will do it, and then the developer to the government. In the bidding of concession project, the government should provide a series of conditions such as "three supplies and one leveling" and grid connection for the project site, and make commitments in the tender documents. As a bidding enterprise, in accordance with the scale, duration and electricity price of the winning bid to implement, scale efficiency and cost reduction has become the inherent nature of the concession project bidding.

Figure 3.5 On October 14, 2003, China's first phase of the wind power concession project signed the Concession Agreement and Power Purchase Agreement.

On September 1, 2003, the first phase of the wind power concession project was opened in Beijing. Jiangsu Rudong Wind Farm (100,000 kW) was won by Huarui Group, and Guangdong Huilai Shibeishan Wind Farm (100,000 kW) was won by Guangdong Energy Group.

In 2004, the second phase of the wind power concession project was opened. The concession project installation capacity of 650,000 kW, began to touch the large-scale development of wind power grid planning and construction and efficiency, cross-provincial sales of wind power and thermal power and other key issues, but also increased the localization rate of wind turbines from 50% to 70%. This is a positive and significant change from the first phase of the wind power concession project.

So far, China has conducted a total of five phases of wind power concession tender, identified 49 wind farm projects, involving Guangdong, Jiangsu, Inner Mongolia, Jilin, Gansu and Hebei and other regions, installation capacity of 8.8 million kilowatts. In addition to accelerating the large-scale development of wind power, it has also accelerated the process of localization of wind power equipment, providing market opportunities for Chinese manufacturers who have just entered the wind power industry.

Figure 3.6 On September 1, 2004, the second phase of the wind power concession project was opened in Beijing, which increased the localization rate of wind turbines from 50% to 70%.

On August 16, 2005, the third phase of the wind power concession project was opened in Beijing. Compared with the first two phases of the concession project, in this tender, the provision that "the bidder who promises the lowest feed-in price is the winning bidder" was cancelled.

The fourth phase of the wind power concession project opened on August 16, 2006 has made a great adjustment to the bid evaluation method, increased the weight of the localized scheme, further reduced the proportion of the bidding price, and made the bidding more rational. Since developers are required to bundle bids with wind turbine manufacturers, participating manufacturers have signed supply agreements with major component suppliers, and formulated plans and schedules to achieve 70% localization, which has promoted the localization process of wind power equipment to a greater extent.

The fifth phase of the wind power concession project was opened on November 30, 2007. The tender used a completely different pricing scheme, whoever bid closer to average tender price will get higher scores, The aim is to change the unreasonable low price bidding scheme of some bidders and guide the bidders to bid rationally.

So far, China has conducted a total of five phases of wind power concession tender, identified 49 wind farm projects, involving Guangdong, Jiangsu, Inner Mongolia, Jilin, Gansu and Hebei and other regions, installation capacity of 8.8 million kilowatts. In addition to accelerating the large-scale development of wind power, it has also accelerated the process of localization of wind power equipment, providing market opportunities for Chinese manufacturers who have just entered the wind power industry.

Looking back, Shi Lishan believes that the concession project bidding policy is very effective, it mobilizes the enthusiasm of the industry, and indeed provides a platform for the domestic manufacturing industry to start through large-scale construction, so that Chinese manufacturing enterprises can grow rapidly. By 2007, China's new wind power market share of domestic products has reached 55%, and the manufacturing capacity of domestic wind turbine equipment has been greatly improved. In terms of wind farm construction, companies have mastered the technology and experience of wind farm operation management, in particular, the reasonable price of electricity was found through the bidding of concession projects, and formed four types of regional benchmark electricity prices in the country.

3.1.3 Move towards Marketization

In order to promote the healthy and sustainable development of wind power, China's electricity price policy has continued to "decline."

The "Renewable Energy Law" clarified the *feed-in tariff*, the full purchase, the whole grid allocation, and the bidding price of the concession project explored the price level under different wind energy resource conditions, and the combination and effect of the two gave rise to four types of zoning electricity price policies based on wind energy resource conditions.

In July 2009, the National Development and Reform Commission issued the *Notice on Improving the on-grid Tariff for Wind Power,* that is, China's first regional feed-in tariff policy for wind power, which determined the principle of formulating onshore wind power benchmarking on-grid tariff by resource area, and divided the country into four types of wind energy resource areas according to wind energy resources and construction conditions. Accordingly, four types of wind power benchmark electricity prices of 0.51 yuan/kWh, 0.54 yuan/kWh, 0.58 yuan/ kWh and 0.61 yuan/ kWh have been formulated. This feed-in tariff policy has established a unified electricity price standard for the development of wind power in China, and has also influenced and standardized the rhythm of wind power development in various places, so that China's wind power has officially entered the mature and large-scale development track.

Resource Zone		China Wind Power Feed-in Tariff (FIT) Evolution (Unit: RMB/kWh)						
		2009-2014	2015	2016-2017	2018	2019	2020	2021
Onshore	Zone I	0.51	0.49	0.47	0.40	0.34	0.29	Grid parity
	Zone II	0.54	0.52	0.50	0.45	0.39	0.34	
	Zone III	0.58	0.56	0.54	0.49	0.43	0.38	
	Zone IV	0.61	0.61	0.60	0.57	0.52	0.47	
Offshore	Intertidal		0.75	0.75	0.75	≤ Onshore benchmark price		No national subsidies
	Offshore	Concession bidding	0.85	0.85	0.85	0.80	0.75	

Figure 3.7 China's wind power feed-in tariff (FIT) phase-down mechanism.

In the following 10 years, in order to promote the healthy and sustainable development of wind power, China's electricity price policy has continued to "decline." On December 31, 2014, the National Development and Reform Commission issued the *Notice on Appropriate Adjustment of Onshore Wind Power Tariff Policy*; In December 2015, the National Development and Reform Commission issued the *Notice on Improving Benchmark On-grid Tariff Policies for Onshore Wind Power*; in December 2016, the National Development and Reform Commission issued the *Notice on Adjusting Benchmark On-grid Tariff for Photovoltaic and Onshore Wind Power*, which repeatedly lowered the feed-in tariff for four types of wind energy resources to 0.40 yuan/kWh, 0.45 yuan/kWh, 0.49 yuan/kWh and 0.57 yuan/kWh in 2018. During this period, in June 2014, the National Development and Reform Commission issued the *Notice on the On-grid Tariff Policy for Offshore Wind Power,* pricing offshore wind power: 0.75 yuan/kWh in the intertidal zone and 0.85 yuan/kWh in the offshore zone.

The end of the wind power feed-in tariff decline is grid parity. In May 2017, the Comprehensive Department of the National Energy Administration issued the *Notice on Launching Demonstration Projects of Wind Power Grid Parity*. Three months later, the National Energy Administration issued the *Notice on Release of the List of Wind Power Grid Parity Demonstration Projects*, announcing a list of 13 grid-parity demonstration projects. On May 21, 2019, the National Development and Reform Commission issued the *Notice on Improving the Pricing Mechanism for Grid-Connected Wind Power*, which clarified the road map for the affordable development of wind power in China. This means that from 2019, China's wind power benchmark feed-in price has been adjusted to the guide price, and its feed-in price level has been determined through competition, and wind power price subsidies have become history.

It was 3 months after the "Notice on the announcement of wind Power parity online Demonstration Project" was issued that CNNC Heyazi 50,000 kW wind power project was connected to the grid. The project is one of the first 13 grid-parity demonstration projects announced by the National Energy Administration in 2017, and is the first grid-parity demonstration project completed in China, marking the first step in China's wind power parity attempt.

By 2021 and 2022, China achieved grid parity for onshore and offshore wind power respectively, reaching price equivalence with the benchmark tariffs for coal-fired power generation.

By 2021 and 2022, China achieved grid parity for onshore and offshore wind power, respectively, reaching price equivalence with the benchmark tariffs for coal-fired power generation. This means the end of the wind power price subsidy policy, and also indicates that this policy has completed its historical mission and has been honorably "retired."

Figure 3.8 On August 14, 2019, the CNNC Heyazi 50,000 kW wind power project was connected to the grid, which became the first fully commissioned grid-parity project in China.

But the Chinese government has not stopped further promoting wind power to the market. In 2024, China's wind power has already fully adapted to the pace of grid-parity development, and the new installation capacity has reached a record high, which has basically formed a good opportunity for wind power to fully enter the market. On February 9, 2025, the National Development and Reform Commission and the National Energy Administration issued the *Notice on Deepening the Market-oriented Reform of New Energy on-grid Electricity Price to Promote the High-quality Development of New Energy*, proposing to adhere to the direction of market-oriented reform, and promote the full entry of new energy on-grid electricity into the power market and the formation of prices through market transactions. The policy clearly requires that the on-grid electricity of new energy projects (wind power, solar power generation) in principle all enter the power market, and the on-grid price is formed through market transactions. It is pointed out that with June 1, 2025, as the time window, distinguish between stock projects and incremental projects, establish a price settlement mechanism for sustainable development of new energy, maintain the cohesion of stock project policies, and stabilize the income expectations of incremental projects. In other words, from June 1, 2025, China's wind power will enter a new period: a full market development stage.

3.1.4 Take Multiple Measures to Increase the Scale

With the promulgation and implementation of the "Renewable Energy Law of the People's Republic of China," the concession projects have accumulated a lot of technology and development experience, and the definition and implementation of the electricity price policy, there are only two challenges to block the large-scale development of wind power in China: first, it is necessary to accurately evaluate their own resources "background" to lay the foundation for the next step of development; second, what is the starting point to leverage the national wind power large-scale development boom.

Wind energy resources are the foundation of wind power project planning and construction, and resource endowment will directly affect the cost and scale of wind power development, and even involve whether China can develop wind power as the "power backbone" described by Qian Xuesen.

As early as 2003, when the development of wind power was just starting, China organized and carried out the evaluation of wind energy resources, mainly based on the experience of the national hydropower resources evaluation. The National Development and Reform Commission arranged part of the funds, and the provincial development and reform commissions arranged supporting funds. Based on the observation data of meteorological stations, the analysis and evaluation of the national wind energy resources were carried out and the national wind energy resources evaluation report was put forward. It provides resource guarantee for the development of wind power.

In 2006, under the guidance of the *Implementation Opinions on Promoting the Development of the Wind Power Industry* and the *Interim Measures for the Management of Special Funds for the Industrialization of wind Power Equipment*, China started the research on wind energy resource evaluation, wind power equipment testing and technical specifications, which effectively improved the technical capability of the development of the service industry.

In 2007, according to the requirements of the Renewable Energy Law, the detailed investigation and evaluation of China's wind energy resources began. The Ministry of Finance arranged funds, led by the Meteorological Administration, set up a national long-term monitoring system for wind energy resources, established a national wind energy resources evaluation system integrating professional observation network, numerical simulation and comprehensive evaluation, and carried out a more comprehensive and systematic evaluation of wind energy resources.

With the progress of wind power technology in recent years, today's wind energy resources are no longer a condition that restricts China's wind power construction. According to the data released by the National Energy Administration and the National Climate Center, by 2024, China's onshore wind power at a height of 140 meters can be developed by 11.4 billion kilowatts, offshore wind power at a height of 150 meters can be developed by 2.8 billion kilowatts, wind energy resources development potential is huge.

China proposed the construction of the "Wind Power Three Gorges" concept, with a loud slogan, to open the prelude to the development of large wind power base.

In the establishment of the national wind energy resources evaluation system at the same time, China proposed the construction of the "Wind Power Three Gorges" concept, with a loud slogan, to open the prelude to the development of large wind power base.

The reason why the country put forward the idea of building "Wind Power Three Gorges" in 2006 is mainly that the "Three Gorges project" has built more than 20 million kilowatts of installation capacity on the basis of flooding a large number of arable land and moving one million people. In China, there are a large number of beaches along the coast and vast Gobi and desert in the west, which can build a large wind power base exceeding the installation capacity of the Three Gorges Power Station without occupying arable land or affecting the population. Therefore, the development idea of "building a large base and integrating into a large power grid" is proposed, and the construction of a large wind power base in Jiuquan, Gansu Province, is taken as the starting point. Construction of a large wind power base of 10 million kilowatts began.

Figure 3.9 In 2009, the Jiuquan 10 GW wind power base was fully launched, marking China's start of building the "Onshore Three Gorges."

In August 2008, the bidding for 3.8 million kilowatts of wind power equipment in the first phase of Jiuquan wind power base (wind power landmark) was completed, marking the construction of Jiuquan wind power base has been fully started. In the same year, the National Development and Reform Commission issued the *11th Five-Year Plan for the Development of Renewable Energy*, proposing the construction of 7 million kilowatt wind power bases, respectively in Jiuquan, Gansu Province, Hami, Xinjiang, Hebei, Jilin, eastern Inner Mongolia, western Inner Mongolia, and coastal Jiangsu.

Figure 3.10 The Jiuquan 10 GW wind power base.

Facts have proved that the proposal and implementation of the idea of building a gigawatt wind power base is a milestone event in the development process of China's wind power. Since then, China has gradually formed a national wind power base construction booming trend, until the realization of parity development, wind power base construction is still a strong starting point to expand the scale of wind power development.

In October 2021, the National Development and Reform Commission, the National Energy Administration, the Ministry of Finance, the Ministry of Natural Resources, the Ministry of Ecology and Environment, the Ministry of Housing and Urban-Rural Development, the Ministry of Agriculture and Rural Affairs, the China Meteorological Administration, and the National Forestry and Grassland Administration issued the *Notice on the Issuing the 14th Five-Year Plan for Renewable Energy Development* proposing to optimize the development mode and develop renewable energy on a large scale, including vigorously promote the localized development of wind power and photovoltaic power generation, actively promote the distributed development of wind power and photovoltaic power generation, and coordinate the integrated development of water-landscape integrated bases. Nine clean energy bases are planned, focusing on the construction of five offshore wind power bases in Shandong Peninsula, Yangtze River Delta, Southern Fujian, East Guangdong and Beibu Gulf.

In the same year, the General Office of the National Development and Reform Commission and the Comprehensive Department of the National Energy Administration issued the *Notice on Issuing the First Batch of Large-Scale Wind-Solar Base Construction Project Portfolios Focused on Desert, Gobi and Wasteland Areas*, layout and promote the development of large-scale wind power photovoltaic bases focusing on desert, Gobi and desert areas. Two months later, the National Development and Reform Commission and the National Energy Administration issued *the Planning and Layout Plan for Large-Scale Wind Power Photovoltaic Bases Focusing on Desert, Gobi and Wasteland Areas*, which plans to focus on Kubuqi, Ulan Buhe, Tengger and Batangjilin deserts, supplemented by other deserts and Gobi areas, and comprehensively consider coal mining subsidence areas to plan and construct large-scale wind power and photovoltaic bases. By 2030, the total installation capacity of scenic base is planned to be about 455 million kilowatts.

By the end of 2024, all 96.9 GW of projects under China's first batch of Desert-Gobi-Wasteland (DGW) wind-solar bases had all connected to the grid, the second batch of 41.83 million kilowatts projects got 13 million kilowatts connected to the grid, the third batch of 93.26 million kilowatts projects got 2.73 million kilowatts connected to the grid.

In order to expand the scale of wind power development according to local conditions, in addition to promoting the development of wind power bases, the relevant competent authorities have sized up the situation and formulated different guidance policies in markets with different objective conditions.

On July 29, 2014, the National Energy Administration issued *the Notice on Strengthening the Relevant Requirements for the Development and Construction Management of wind power Projects*, encouraging the construction of decentralized access wind power projects, and proposed that the energy authorities of provinces (regions and municipalities) organize the preparation of regional decentralized wind power development plans, and incorporate them into the annual implementation plan management.

Subsequently, low-wind speed wind power technology gradually mature, with the *13th Five-Year Plan for Wind Power*

Development proposed to accelerate the development of onshore wind energy resources in central and eastern and southern regions, provinces have prepared decentralized wind power development plans, decentralized wind power gradually achieve large-scale development. On April 16, 2018, the National Energy Administration issued the "Interim Management Measures for the Development and Construction of Decentralized Wind Power Projects," improving the management process and working mechanism of decentralized wind power.

2012–2022 was also a critical period for the development of offshore wind power. On December 8, 2014, the National Energy Administration issued *the Notice on Issuing the National Offshore Wind Power Development and Construction Plan (2014–2016)*, approving the construction of a number of offshore wind power projects.

On December 29, 2016, the National Energy Administration and the State Oceanic Administration issued the *Offshore Wind Power Development and Construction Management Measures*, which clarified the offshore wind power management process, and standardized the development and construction order, and proposed that the provincial and below energy authorities in accordance with relevant laws and regulations, according to the offshore wind power development plan approved by the National Energy Administration, approved offshore wind power projects with construction conditions. Subsequently, all coastal areas completed the resource survey and compiled the development plans for offshore wind power in provinces (autonomous regions, municipalities directly under the Central Government) and cities and counties, and vigorously promoted the large-scale development of offshore wind power. In addition, the "Circular on the issuance of the 14th Five-Year Plan for Renewable Energy Development" further pointed out that in industrial parks, economic development zones, oil and gas mining areas and surrounding areas, actively promote the decentralized development of wind power; innovate wind power investment and construction models and land use mechanisms, implement the *Thousands of Towns and Villages Wind-Driving Action*, and vigorously promote rural wind power development. Among them, the *Thousands of Towns and Villages Wind-Driving Action* will vigorously promote the construction of rural wind power in the county as a unit, and promote the development of rural wind power in about 100 counties and 10,000 administrative villages.

3.1.5 The Right Development Track

> *According to the "Notice on the Requirements of the National Energy Administration on Regulating the market order of wind power Equipment," when wind project developers conduct equipment procurement bidding, Type Certification products should be explicitly required. Units without Type Certification are not allowed to participate in the tender.*

If a jade is not cut and polished, it cannot be made into anything. With the development of more and more wind power projects and the rapid expansion of wind power development scale, China's wind power has gradually revealed some aspects that need to be improved and optimized in the development and manufacturing links.

On the one hand, China's wind power project approval mechanism needs to keep up with the pace of large-scale development. With the construction of wind power in full swing across the country, according to the management of the national construction of wind power projects, on August 25, 2011, the National Energy Administration issued the *Interim Measures for the Management of Wind Power Development and Construction*, the approval of wind power projects below 50,000 kW was delegated to the provincial Development and Reform Commission, simplifying the wind power project construction management procedures.

On May 15, 2015, the National Energy Administration issued the *Notice on Further Improving the Management of Wind Power Annual Development Plans*, implementing the overall requirements of the State Council for streamlining administration and delegating power to lower levels and transforming government functions, further simplifying approval procedures, clarifying responsibilities at all levels, and proposing to no longer uniformly issue approval plans containing project lists. Areas that do not have abandoned wind independently propose the scale of development, and strengthen and improve the supervision during and after the event.

On the other hand, with the increasingly fierce competition in the wind power equipment market, after 2010, the problem of "price war" has appeared in China's wind power manufacturing side. The price war not only led to a sharp decline in the price of wind turbines, the industry fell into losses, but also increased quality accidents, and industrial development entered the trough.

In November 2011, the National Energy Administration issued the *Notice on Strengthening the Relevant Requirements for Wind farm Safety Management*, requiring the strengthening of wind farm safety management, and carry out wind power accident information reporting work, so that the relevant departments to grasp the wind farm safety situation in a timely manner, take appropriate measures to deal with and supervise. On May 23, 2013, the National Energy Administration further issued the *Notice on Strengthening the Construction of the Wind Power Industry Monitoring and Evaluation System*, aiming to comprehensively and accurately grasp the development information and situation of the wind power industry, improve the industrial technology level and engineering quality, and promote the healthy and sustainable development of the wind power industry.

According to the *Notice on the Requirements of the National Energy Administration on Regulating the Market Order of Wind Power Equipment*, when wind power development enterprises conduct equipment procurement bidding, Type Certification products should be explicitly required. Wind turbines without Type Certification are not allowed to participate in the tender.

On September 5, 2014, the National Energy Administration issued the *Notice on the Requirements of the National Energy Administration on Regulating the Market Order of Wind Power Equipment*, which pointed out that "the quality of wind power equipment is an important basis for the sustainable and healthy development of the wind power industry, and the testing and certification system is an important measure to ensure the quality of equipment," and proposed that when wind power development enterprises conduct equipment procurement bidding, Type Certification products should be explicitly required. Wind turbines without Type Certification are not allowed to participate in the tender.

The release of a series of policies has led China's wind power to the right development track, coupled with the efforts of the entire industry, to achieve such a large-scale market today.

3.1.6 On a New Journey

> *Xi Jinping, the president of the People's Republic of China announced that "China will increase its nationally determined contributions, adopt more powerful policies and measures, strive to peak its carbon dioxide emissions before 2030, and strive to achieve carbon neutrality by 2060."*

On 22 September 2020, Xi Jinping, the president of the People's Republic of China announced to the world at the general debate of the seventy-sixth session of the United Nations General Assembly, "China will increase its nationally determined contributions, adopt more powerful policies and measures, strive to peak its carbon dioxide emissions before 2030, and strive to achieve carbon neutrality by 2060."

On December 12, 2020, at the Climate Ambition Summit, President Xi delivered an important speech entitled "Building on Past Achievements and Opening up a New Journey for the Global Response to Climate Change" through video, announcing a series of new measures for China's nationally determined contributions. He pointed out that by 2030, China's carbon dioxide emissions per unit of GDP will be more than 65 percent lower than in 2005, non-fossil energy will account for about 25 percent of primary energy consumption, forest stock will increase by 6 billion cubic meters over 2005, and the total installation capacity of wind and solar power will reach more than 1.2 billion kilowatts.

The proposal and promotion of the carbon peaking and carbon neutrality goals has become a common task for all walks of life in China in the next few decades, which has injected new vitality into the development of wind power and put forward new requirements, and wind power is about to usher in a new phase of development.

From October 14 to 16, 2020, at the Beijing Wind Power 2020 (CWP 2020) with the theme of "Leading a Green Recovery for a Better Future," representatives from more than 400 wind energy companies around the world unanimously adopted and jointly released the *Wind Energy Beijing Declaration*, It aims to comply with the latest requirements of the national strategy to address climate change, promote the formulation of more active industrial policies, promote the wind power industry to maintain high-quality development, and lay a solid foundation for the green recovery of the global economy in the post-epidemic era and the construction of a zero-carbon society. The declaration proposes that, considering the practical feasibility of resource potential, technological progress trend, grid-connected consumption conditions and so on, in order to achieve the purpose of starting to connect with the carbon neutral goal, in the "14th Five-Year Plan," it is necessary to set a development space for wind power that is compatible with the national strategy of carbon neutrality—an annual increase of more than 50 million kilowatts of installation capacity. After 2025, China's annual installation capacity of wind power should be no less than 60 million kilowatts, at least 800 million kilowatts by 2030, and at least 3 billion kilowatts by 2060.

The proposal of national and industry development goals has provided a strong driving force for the development of China's wind power industry. Through the efforts of the whole industry, China's wind power has been ahead of schedule or exceeded part of the short- and medium-term goals. For example, by 2024, China has completed the goal of "total installation capacity of wind power and solar power generation will reach more than 1.2 billion kilowatts" 6 years ahead of schedule. The new installation capacity of wind power in 2023 and 2024 also far exceeds 50 million kilowatts, reaching 79.37 million kilowatts and 86.99 million kilowatts, respectively.

Figure 3.11 On October 14, 2020, representatives from more than 400 wind energy companies around the world unanimously adopted and jointly released the "Wind Energy Beijing Declaration."

Figure 3.12 In February 2023, Shi Lishan (second from the right) was presented with the "Wind Power Pioneer · Outstanding Contributor Award"; the person first from the left is Qin Haiyan, the author of this chapter.

On February 26, 2023, the Chinese Wind Energy Association (CWEA) convened the "2023 New Year Reception of China Wind Power Industry" in Beijing, where Shi Lishan was presented with the "Wind Power Pioneer · Outstanding Contributor Award" in recognition of his pivotal role in advancing China's wind power development. The citation read by veteran wind power expert Shi Pengfei proclaimed: "You forged the legal bedrock for high-quality industry growth, ignited the policy engine driving gigawatt-scale wind development, refined the market-driven governance that steers our sector's course, exemplified the innovative, enterprising spirit that defines our core values. Two decades on, undaunted by hair graying at the temples, you charted strategic blueprints guiding wind pioneers' journey. Now gazing back at verdant mountains and lucid waters, eight thousand leagues bear witness to renewables' mighty flame."

3.2 Keep the Initiative in Our Own Hands

From unit capacity of 10 kW to 26 MW, from the rotor diameter of 10 meters to 300 meters, from the tower height of 20 meters to 190 meters, from the altitude of 5200 meters to –120 meters, from all relying on imports to parts localization rate of 99%, China's wind power has realized its dreams and created too many miracles through independent innovation.

In November 1986, Wang Wenqi, director of Xinjiang Institute of Water Resources and Hydropower, led several young people to erect two wind turbines from Denmark on the bank of Chaiwopu Lake in Dabancheng, one is Wincon-100W grid-connected wind turbine, and the other is Wincon-55kW independent operation wind turbine.

A year later, at the age of 29, Wu Gang switched from being a school teacher to taking over as the head of the Chaiwopu wind power test site. At the Dabancheng Chaiwopu wind power test site, Wu Gang's dream is to develop wind power in plots like sorghum. Later, this dream was painted into a large oil painting, almost hung all over a wall. When Chinese and foreign experts visited and conducted research at Dabancheng Wind Farm, most of them would take photos in front of the painting as a memento.

Figure 3.13 October 1989, Dabancheng Wind Farm I in Xinjiang was commissioned. Funded by a grant from the Danish government, the project installed 13 Bonus 150 kW wind turbines and 1 Wincon 100 kW wind turbine, with a total installation capacity of 1050 kW, making it the largest in Asia at that time. Later, eight turbines from three manufacturers were introduced through Germany's "Gold Program," including 2 Tacke 600 kW turbines, which had the largest single-unit capacity in China when commissioned in 1996.

Figure 3.14 Winter 1993, Wu Gang (left) with international experts at Dabancheng Wind Farm I.

On October 24, 1989, in Dabancheng, the wind farm of Wu Gang's dream was connected to the grid and commissioned. At that time, they utilized the Danish government's $3.2 million mixed loan to import 13 units of 150 kW stall-regulated turbines from Denmark's Bonus company. Along with the Wincon–100 kW turbines originally relocated from the shores of Chaiwopu Lake, Dabancheng Wind Farm has a grid-connected capacity of 2050 kW, becoming the largest wind farm in terms of installed capacity in China and even in Asia at that time.

Figure 3.15 *1989, Central control room of Dabancheng Wind Farm.*

In the 20 years from 1989 to 2009, Dabancheng Wind Farm has commissioned more than 300 domestic and overseas units of various types, involving more than 30 models from more than 10 manufacturers. Single capacity from tens of kilowatts, hundreds of kilowatts to thousands of kilowatts, become the base camp to explore China's wind power development road.

Later, Xinjiang Electric Power Bureau organized personnel to visit Denmark and purchased eight wind turbines, including four Bonus Company's 300 kW turbines and four Nordtank company's 300 kW turbines. The reason for choosing products from two enterprises was that when undertaking this project, Xinjiang

Electric Power Bureau had already taken into account the issue of wind turbine localization: components that could be produced in-house would not be imported, and the two models were compared and evaluated against each other, laying the foundation for wind turbine localization.

In the 20 years from 1989 to 2009, Dabancheng Wind Farm has commissioned more than 300 domestic and overseas wind turbines of various types, involving more than 30 models from more than 10 manufacturers. Single capacity from tens of kilowatts, hundreds of kilowatts to thousands of kilowatts, become the base camp to explore China's wind power development road, which precipitated the formation of a museum that reflects the history of China and the world's wind power relationship.

Figure 3.16 Wugang, chairman of Goldwind Technology.

In 2002, Wu Gang, who has been carrying the dream of developing wind power and realizing the localization of wind turbines, began to serve as the chairman of Goldwind Technology, the world's largest annual installed wind power machine manufacturer, in 2002. His dedication to wind power, from Dabancheng to Goldwind Technology, and to the world.

3.2.1 First Attempt

Dabancheng is a landmark that can represent the early development of China's wind power, but it is not the only one that can be called the birthplace of China's wind power technology and industry. Before foreign models were introduced into China, many Chinese enterprises and scientific research institutions were independently developing wind turbines.

In the late 1950s, China started the research of modern wind power technology. From 1957 to 1959, in Jiangsu, Jilin, Liaoning, Anhui, Xinjiang and other places, some small wind turbines with a power of less than 10 kW and a wind rotor diameter of less than 10 m were installed.

In July 1972, Shaoxing Grain and Oil Machinery Processing Factory in Zhejiang Province produced a prototype called FD13-18 kW wind turbine. The prototype was jointly designed by East China Electric Power Design Institute, Zhejiang Machinery Research Institute, Zhejiang Hydropower Design Institute and other institutions, and was installed in the Xiong'e Peak in the mountains near Shaoxing. It ran intermittently for nearly 300 hours before being shut down due to malfunction.

Figure 3.17 July 1972, Shaoxing Grain and Oil Machinery Processing Factory in Zhejiang Province trial-produced the FD13-18 kW wind turbine.

The first Chinese wind turbine to be finalized and put into mass production is model FD1.5-100W.

During this period, most of China's wind turbines exist in the form of prototypes or small batch products, and it is difficult to form mass production and application results. Until 1975, according to the tasks assigned by the Ministry of Agriculture and Forestry, the Grassland Research Institute of the Chinese Academy of Agricultural Sciences undertook the research and development of small wind power machines for the service objects of yurts in pastoral areas. In 1975, two prototype FG-1.3 wind turbines were designed and trial-produced. In 1976, the first batch of 60 wind turbines was produced. From 1976 to 1977, the Department of Electrical Engineering of Tsinghua University cooperated with Grassland Research Institute and Shangdu Agricultural Machinery Factory to improve the FG-1.3 wind turbines and develop the second-generation FG-1.5 model. The model was certified in May 1977 and became the first Chinese wind turbine to be finalized and put into mass production, model name FD1.5-100W.

In May 1982, the FDS21-55 kW wind turbine jointly designed by Fujian Electric Power Center Test Institute and Fujian Machinery Research Institute was successfully connected to the grid in Fujian Pingtan Wind Power Test Station. The development of this wind turbine was included in the national "6th Five-Year Plan (1981–1985)" science and technology plan.

Figure 3.18 May 1982, the FDS21-55 kW wind turbine (developed in 1978) was grid-connected in Pingtan, Fujian.

In 1988, as an important exploration in the development process of China's wind power equipment, the first 200 kW wind turbine FD32-200 jointly developed by Hangzhou Machinery Design and Research Institute, Fujian Machinery Factory, Shanghai FRP Research Institute, Fujian Electric Power Experimental Research, Pingtan County wind Power generation Test Station and other institutions was hoisted in Fujian Pingtan Wind Farm. The wind turbine was connected to the grid in December 1991 and passed the technical appraisal organized by the Ministry of Energy in April 1993.

According to this time node to compare, China's independent research and development of wind turbines in unit capacity, has been basically maintained at the same level with the products of international manufacturers.

In order to better study and test wind power equipment, the Railway Power Bureau of the Ministry of Water Resources and Electric Power established the Badaling wind power test station in 1979, and transferred it to the Beijing Electric Power Science Institute of North China Electric Power Administration in June 1983. According to the requirements of the state at that time, the station will gradually be built into a nationwide wind power research and test center, technical training center, science and technology information center and technology promotion center. Taking advantage of better wind energy resources and superior geographical conditions, the test station has installed dozens of wind turbines of various types and models, established a 104 m wind tower and various test equipment, and conducted a large number of scientific experiments, which has played an important role in demonstrating and promoting the development of China's wind power industry.

In addition to new wind turbines and test stations to test new models, China had already built a "commercial demonstration" wind farm before Dabancheng Wind Farm was completed. In April 1986, the Shandong Rongcheng Malan wind farm, jointly invested by the Shandong Provincial government and the Ministry of Aviation Industry, installed three Vestas V15-55/11 kW wind turbines and connected to the grid in May of the same year. This was the world's most advanced wind turbine at that time, just like the "enlightenment teacher" of China's wind power, standing on the coast of China's earliest rising sun. In the 23rd year of operation of the wind farm, Vestas mobilized global resources to do a deep overhaul for it free of charge, replacing some of the blades, until 2015 was honorably retired.

Figure 3.19 May 1986, three Vestas 11 kW wind turbines installed at Rongcheng Malan Wind Farm were grid-connected.

Malan wind farm is the originator of wind power commercialization in China, and it is also the "holy land" for wind power strivers to visit and learn. After the wind farm was completed and connected to the grid, it has received 18 investigation teams from Hainan, Inner Mongolia, Guangdong and other places, 37 groups of experts and scholars from universities, research institutes and related institutions, and more than 30 national ministries, provinces and prefectural leaders. Even in the late 1990s, Shandong Rongcheng's Mulan Wind Farm remained a pilgrimage site for China's wind pioneers.

3.2.2 Walk on Two Legs

When unit capacity of advanced wind turbines at home and abroad has developed to 600 kW, China's wind power technology innovation has appeared two parallel routes.

In 1993, Wu Gang became the deputy general manager of Xinjiang Wind Energy Company. In terms of organizational relations, Xinjiang

Wind Energy Company is a subsidiary company of Xinjiang Water Resources Department, and Dabancheng Wind Farm is a property of Xinjiang Wind Energy Company. Because Xinjiang Electric Power Bureau also built a wind farm in Dabancheng, Xinjiang Wind Power Company's wind farm was changed to Dabancheng Wind Farm I.

Even in difficult days, Wu Gang also led the team to adhere to the study and research of wind power technology, expand international communication, and study the operation of 14 wind turbines of Dabancheng Wind Farm became the most important point at that time. In 1995, Xinjiang Wind Energy Company welcomed a ray of spring light from Germany's "Gold Program." Xinjiang Wind Energy Company used the "Gold Program" to introduce eight large wind turbines from three German manufacturers, including Tacke and Jacobs, and the cumulative installation capacity of Dabancheng Wind Farm I reached 6,100 kW.

The "Gold Program" was launched by the German government in 1995–1996 to support the construction of wind power and other new energy projects in developing countries by providing grant assistance of two-thirds of the equipment price. China has a total of six projects in wind power. In addition to the Dabancheng Wind Farm I, the Zhurihe wind farm in Inner Mongolia has also introduced the installation of four German HSM250T wind turbines through the use of a partial grant from Germany's "Gold Program."

Dabancheng Wind Farm I introduced a new type of wind turbine through the "Gold Program," and the Dabancheng Wind Farm II, which was invested by Xinjiang Power Bureau, was expanded by using the "Dual-Intensification Loan Program" of the State Economic and Trade Commission around 1996. This increased the installation capacity of the wind farm from 10,100 kW to 57,500 kW by the end of 1998, installed scale and unit capacity ranked first in the country.

The "Dual-Intensification Loan Program" was organized by the State Economic and Trade Commission in 1996, for the purpose of "accelerating technological transformation and increasing investment," in accordance with the principle of supporting strength and excellence, discount loans were provided, and four wind farms, namely Dabancheng Wind Farm II, Huiteng Xile Wind

Farm, Kuocangshan Wind Farm and Zhangbei Wind Farm, were selected for key transformation. Through this project, 133 units of 600 kW wind turbines, 13 units of 300 kW wind turbines, a total of 83,700 kW, were completed and accepted in 1997 and 1998.

When unit capacity of advanced wind turbines at home and abroad has developed to 600 kW, China's wind power technology innovation has appeared two parallel routes.

The first route is to introduce wind turbine design and process technology from international wind turbine enterprises through technical trade cooperation, and manufacture joint venture brand products in China.

At the beginning of 1996, Beijing Wandian Co., LTD., which is affiliated to the Aerospace Industry Corporation and specializes in the design and development of large wind turbines, signed a contract with the Austrian PEHR company to introduce the design and process technology for the manufacture of 600 kW variable pitch type turbines, and began to develop and produce them. The first prototype was put into trial operation in Huiteng Xile Wind Farm in Inner Mongolia in January 2000. In April 2003, three more prototypes were connected to the grid at Xilinhot Wind Farm in Inner Mongolia.

Figure 3.20 Signing ceremony for localized wind turbines under the "Embracing the Wind Plan."

In March 1996, the former State Planning Commission launched the "Embracing the Wind Plan," which aims to establish joint ventures with foreign manufacturers in the way of combining technology and trade, introduce technology, digest and absorb, and achieve the ability of independent development, design and manufacture of large wind turbines. "Embracing the Wind Plan" adopts the way of bidding and evaluation to determine Xi'an Aero Engine Group Company and China First Tractor Construction Machinery (Group) Company as large wind turbine final assembly plants. In 1998, Xi'an Aero Engine Group Company and Germany Nordex company jointly established Xi'an Weide Wind Power Equipment Co., LTD., China First Tractor Construction Machinery (Group) company and Spain MADE recycling technology company jointly established Yito-Meide (Luoyang) Wind Power Equipment Co., LTD. In September 2000, Xi'an Vide N43/600 kW wind turbine prototype was commissioned in Xianren Island Wind Farm, Yingkou, Liaoning Province. In October of the same year, Yituo-Meide 660 kW wind turbine prototype was also commissioned in the wind farm.

The second route is through the introduction, digestion, absorption, re-innovation, through the introduction of foreign technology and promote the localization of products, and to develop domestic brand products.

According to Wu Gang, with the implementation of the "Gold Program," he had the opportunity to lead a team to Germany's Tacke, Jacobs and other companies for technical training and inspection of wind turbine manufacturing. They also visited the Risø National Laboratory in Denmark and were deeply impressed by the experimental data and projects of this laboratory. Like experts from the United States, the United Kingdom and other countries who come to Risø laboratory to study and solve problems, then face new problems, and bring back those questions, Wu Gang has become a "day graduate student" in this laboratory, and every time they come here, they will memorize reports and data. Time after time of technical experience and continuous improvement of cognition, the dream of making wind turbines is more and more solid.

For three years, from 1998 to 2000, Xinfeng Company has developed and commissioned 10 units of 600 kW wind turbines, the first unit of the localization rate is 33%, and to the 10th unit, its localization rate reached 96%.

In the summer of 1997, Xinjiang Wind Energy Company signed a production license and technology transfer contract with Jacobs Company of Germany for stall-controlled 600 kW wind turbines. In 1998, Xinjiang Xinfeng Industry and Trade Co., LTD., the predecessor of Goldwind Technology, was established to promote the localization of 600 kW wind turbines. Wu Gang and colleagues have written more than 200 invitations for cooperation and sent them to more than 200 enterprises and institutions at home and abroad, inviting them to jointly develop parts manufacturing processes and other projects, which are integrated and assembled by Xinfeng Company. After more than one year of hard work, the first 600 kW domestic wind turbine was connected to the grid in Dabancheng Wind Farm I.

Figure 3.21 1998, the first batch of domestic 600 kW wind turbines were grid-connected in Dabancheng Wind Farm, Xinjiang, with a localization rate exceeding 53%.

For 3 years, from 1998 to 2000, Xinfeng Company developed and commissioned 10 units of 600 kW wind turbines; the first unit's localization rate was 33%, and for the 10th unit, its localization rate reached 96%.

Figure 3.22 1999, the national "9th Five-Year" key scientific and technological project—600 kW localized wind turbines successfully passed appraisal and acceptance, marking a solid step for China's wind power in independent intellectual property rights.

Figure 3.23 2002, the national "10th Five-Year" key scientific and technological project arranged the development of 750 kW stall-controlled wind turbines, undertaken by Zhejiang Yunda and Goldwind Technology. The products were commissioned in 2004.

Subsequently, the national "10th Five-Year Plan" scientific and technological research plan arranged the development of 750 kW stall-controlled wind turbine, and Zhejiang Yunda and Goldwind Technology undertook the project. The successful products of the two were commissioned in 2004 and achieved industrial production, which has accumulated valuable experience for exploring the localization and commercialization of China's large-scale wind turbines.

3.2.3 Megawatt Breakthrough

After the batch of 600 kW and 750 kW models, the target model developed by Goldwind Technology is a megawatt-level active stall-controlled type unit, but after conceptual design, load calculation and other links, they found that this type of model is not an ideal technical direction. With the heart of the yearning for innovation and challenge, Wu Gang, bursts the idea of developing other models.

Coincidentally, Wu Gang learned of international R&D on direct-drive permanent magnet generators. This technical route could revolutionize wind turbines by eliminating high-speed mechanical components, using power electronics for grid synchronization. The challenge was the immaturity of high-power power electronics at the time—but given their rapid advancement, this hurdle was confidently addressable.

At that time, foreign countries were only in the research stage, and there was no formed direct-drive permanent magnet prototype, which made Wu Gang more firm in the direction of developing direct-drive wind turbines. This idea has been supported by the national "863" program, so Goldwind Technology has adjusted the development direction of megawatt wind turbines from active stall-controlled type to permanent magnet direct drive.

In 2004, Goldwind Technology and Germany's Vensys company began to jointly design and in June 2005 completed the development of 1.2 MW direct-drive prototype, installed in Dabancheng Wind Farm I, which is China's first local production of megawatt wind turbine. Wu Gang was so bullish on direct-drive technology that it got ahead of the game, buying a 70% stake in Vensys in January 2008.

In fact, the national "863" program not only supports the development of direct-drive permanent magnet wind turbines of Goldwind Technology, but also supports 1.5 MW double-fed wind turbines of the Wind Energy Research Institute of Shenyang University of Technology.

From 2005 to 2008, China's first domestic megawatt wind turbines were launched. In September 2006, Sinovel Wind Power's first 1.5 MW double-fed wind turbine was connected to the grid, and the first 1.5 MW double-fed wind turbines of Dongqi came off the production line. Subsequently, Electric Wind Power, Yunda Shares, Mingyang Smart, United Power, Shenyang University of Technology, CSSC Haizhuang, Jiangsu Xinyu, CNR (CRRC Shandong), and Xiangdian Wind Power (Xinglan Wind Power) have been offline hoisting 1.25~2 MW double-fed wind turbines.

Figure 3.24 January 8, 2008, the first phase of Beijing Guanting Wind Farm was grid-connected.

On January 8, 2008, the first phase of Beijing Guanting Wind Farm was connected to the grid. The project is Beijing 2008 "Green Olympics" project, Beijing's first large-scale wind farm, a total capacity of 50,000 kW, a total of 33 Goldwind Technology 1.5 MW wind turbines installed. The operation of the wind farm provides green guarantee for 20% of the power supply of the Beijing Olympic venues, creating a benchmark for green Olympic supporting projects, and also showcasing the latest achievements of China's wind power technology innovation to the world.

3.2.4 Going into a Technological "No Man's Land"

> *If the Chinese wind power technology was once developed on the shoulders of giants, then from this moment on, China's wind power in technological innovation has entered the "no man's land" without the footsteps of predecessors.*

In December 2022, one kind of wind turbine product shocked the world. The GWH252-16MW offshore wind turbine, jointly developed by Goldwind Technology and Three Gorges Group, rolled off the production line in Fujian Three Gorges Offshore Wind Power International Industrial Park. This model's unit capacity reached an unprecedented 16 MW, setting a new record for the world's largest unit capacity, the world's largest rotor diameter, and the lightest weight per MW.

Figure 3.25 2022, Goldwind Technology's GWH252-16MW offshore wind turbine successfully rolled off the production line in Fujian, setting a new benchmark for global offshore wind power equipment development.

For China's wind power industry, the significance of this unit offline is not only its huge capacity, but more importantly, it is the first time that China's wind power technology and products have reached the world's advanced level. If the Chinese wind power technology was once developed on the shoulders of giants, then from this moment on, China's wind power in technological innovation has entered the "no man's land" without the footsteps of predecessors. In the future, China must rely on its own ability and experience to open up a wider world.

It took China 14 years to go from fully developing its first offshore wind turbine to rolling off the production line of the world's largest model.

Figure 3.26 2010, China's first national offshore wind power demonstration project and Asia's first offshore wind farm—Donghai Bridge Offshore Wind Farm—was commissioned.

In May 2008, the National Development and Reform Commission approved China's first large-scale offshore wind power project, the Donghai Bridge 100 MW offshore wind power demonstration project. The project plans to use offshore wind turbines developed entirely by China. In September 2009, the first three Sinovel Wind 3 MW offshore wind turbines were connected to the grid in this project, marking a significant progress in the development of

China's first offshore wind turbine. On June 8, 2010, the 100 MW offshore wind power demonstration project of Donghai Bridge was connected to the grid, and China's offshore wind power industry took the first step steadily.

The Donghai Bridge 100 MW offshore wind power demonstration project not only uses a purely domestic offshore wind turbine, but also realizes two "first": it adopts the world's first multi-pile cap foundation design; pioneering integrated turbine installation via jack-up vessel with dual cranes.

On September 28, 2010, Longyuan Rudong 30,000 kW offshore (intertidal) test project was completed and commissioned, which not only achieved the breakthrough of zero global offshore intertidal wind farms, but also provided a platform for China's domestic offshore wind turbines to verify technology and operation quality. The project has installed 16 units of 9 models from 8 manufacturers, the actual installation capacity reached 32 MW. In this project, a variety of foundation forms such as concrete cap, multi-pile jacket and single pile without a transition section were also tried.

On September 6, 2011, China's first 5 MW offshore wind turbine was successfully installed in the second phase of Shanghai Donghai Bridge Offshore Wind Farm project, and was commissioned the following year. At this point, the single capacity and rotor diameter of China's offshore wind turbines have begun to develop towards large-scale development.

In July 2018, all wind turbines of Three Gorges Xinghua Bay Sea test wind farm in Fuqing were commissioned, and 14 wind turbines from 8 manufacturers were installed in the project, including 2 Goldwind 154-6.7MW, 2 Taiyuan Heavy Industry 153-5MW, 2 CSIC HZ Windpower 128-5MW, 3 GE150-6MW, 1 Shanghai Electric 154-6MW, 1 Dongfang Electric Wind Power 140-5MW, 1 XEMC Wind Power 140-5MW, 2 Mingyang Smart 155-5.5MW, The total installation capacity was 77.4 MW. In the second phase of the project, which was commissioned 2 years later, one Goldwind 175-8.0MW prototype, one Mingyang Smart 180-8.3MW prototype, and one Dongfang Electric 185-10MW prototype were installed to continue the test and verification of high-power offshore models.

Figure 3.27 July 12, 2020, China's first 10 MW offshore wind turbine was successfully grid-connected in the second phase of Xinghua Bay Offshore Wind Farm (Fuqing, Fujian) owned by China Three Gorges Corporation.

It can be said that the Three Gorges Fuqing Xinghua Bay offshore test wind farm is the arena for large-scale offshore wind turbines. Through the same competition, the best models suitable for Marine geology and wind conditions in Fujian are selected, which not only provides validated models and engineering support for the centralized large-scale development of offshore wind power in Fujian, but also adds new impetus to promote the overall development of offshore wind power in China.

After the removal of all state subsidies for China's offshore wind power projects in 2020, the trend of large-scale offshore wind power models is more obvious. CSIC HZ Windpower installed independently developed H260-18MW offshore wind turbine, and Goldwind Technology GWH252-16MW offshore wind turbine was also completed in December 2022, the former capacity reached 18 MW. Since then, Chinese wind turbine enterprises such as Envision Energy, Mingyang Smart, Yunda Wind Power, Electric Wind Power, Dongfang Electric, CRRC Zhuzhou Institute have offline or tested offshore wind turbines with unit capacity of more than 16 MW. As of March 2025, the largest wind turbine offline in China is Dongfang Electric's 26 MW model, setting a new global record.

Figure 3.28 October 2024, Dongfang Electric's 26 MW wind turbine rolled off the production line.

3.2.5 Open a Bigger Market

> *This is the first low-wind speed wind farm built and commissioned in China, ending the history of no wind power in low-wind speed areas, which means that low-wind speed areas accounting for more than 60% of China's wind energy resources have development value from now on, changing the development pattern of China's wind power industry.*

While stepping up the research and development of wind turbines with independent intellectual property rights, suitable for wind turbines and offshore wind turbines in the "Three North" medium-high wind speed areas, China's wind power technology innovation has been blooming at many points, and models with special adaptability have been developed for different wind energy resources and terrain conditions. This mainly includes three aspects:

First, the development and application of low-wind speed wind power technology has provided a broader space for wind power development.

In 2009, due to the bottleneck of power grid delivery and the lack of local consumption demand, China's "Three North" region experienced "wind curtailment," and the wind power developers were forced to slow down or even stop the development of high-quality wind energy resources in areas.

As a pioneer in the large-scale development of wind power in China, Longyuan Power hopes to realize the large-scale development strategy transfer of local development and local use through the development of low-wind-speed wind power. As a result, the company transferred backbone forces from neighboring Jiangsu Province and other places to field exploration in Lai 'an of Anhui province, accelerated the feasibility study of the project, and repeatedly discussed project solutions with Envision Energy and United Power, and finally came up with a development plan that can meet the internal rate of return (IRR) 10% investment requirements.

On January 6, 2011, Longyuan Anhui Lai'an low-wind-speed wind farm with an average annual wind speed of 5.7 m/s was connected to the grid. This is the first low-wind speed wind farm built and commissioned in China, ending the history of no wind power in low-wind speed areas, which means that low-wind speed areas accounting for more than 60% of China's wind energy resources have development value from now on, changing the development pattern of China's wind power industry.

The success of Lai 'an low-wind speed project mainly depends on two points. First, customized low-wind-speed models suitable for operation in low-wind-speed areas and efficient control strategies; the second is the fine micro-location, which is refined to determine the aircraft position according to the terrain and landform or even the dense or sparse vegetation, so that each unit can capture more wind energy.

In the fall of 2015, an expert from the International Electrotechnical Commission's Renewable Energy Equipment Certification Mutual Recognition System (IECRE) went to China to investigate the development of wind power technology and said with emotion: "China's low-wind speed wind power is a great initiative, allowing the world to see a broader space for wind power development."

Second, the high tower makes the wind energy resources in the high-shear area have the development value.

From the perspective of wind resources, the wind speed changes with vertical distance, which is called wind shear in atmospheric science. In the wind power industry, wind shear is usually used to represent the change rate of wind speed in the vertical direction. If the wind shear index is large, the wind speed increases rapidly with height, that is, the wind speed gradient is large. A small wind shear index means that the wind speed increases slowly with height, that is, the wind speed gradient is small. High shear means that wind speed increases rapidly with height, and higher wind speed can be used after the tower is raised, so that wind power projects in areas with high shear have economic development value or obtain higher returns.

In November 2016, State Power Investment Corporation Limited (SPIC)'s Jiangsu coastal land wind farm III was commissioned, and the installation of the wind turbine tower was raised from the traditional 90–110 meters to 120 meters, and the 121-diameter low-wind-speed model was adopted; it was the first wind farm in China to adopt the high tower.

Figure 3.29 190-meter hybrid wind tower developed by Envision Energy.

The Three Gorges New Energy invested and built China's first floating offshore wind power platform, carrying the world's first typhoon-resistant floating offshore wind turbine to form the "Three Gorges Leading," which successfully connected to the grid in Yangjiang offshore wind Farm in Guangdong Province.

On May 29, 2017, China's first tower of 140 meters was located in Lankao, Henan, and the development of low-wind-speed and high-shear wind resources in the plain area was opened. In September 2020, 160-meter tall wind tower models were connected to the grid in Juancheng, Shandong, refreshing the height record of Chinese wind turbines. A year later, the wind turbine with 170-meter-tall tower was commissioned in Jiaozhou, Shandong, and the record was renewed. By December 2024, the 190 meters mixed tower developed by Envision Energy has been commissioned in Inner Mongolia, which is the latest record for the height of the global mixed tower.

Third, floating wind turbines support the future development of China's wind power to the far-reaching sea.

Far-reaching sea area has rich wind energy resources, which means that far-reaching sea wind power in China and even the world's development potential is huge. According to the World Bank (WB), there are more than 71 billion kilowatts of offshore wind energy available globally, of which the deep sea accounts for more than 70%, but the current utilization rate of these resources is less than 0.5%. According to the latest assessment results of the National Climate Center, China's profound sea wind energy resources can be developed more than 1.2 billion kilowatts.

In order to achieve a breakthrough in far-reaching sea wind power technology and continue to reduce the development cost of far-reaching sea wind power, after 2020, China's wind power development enterprises, machine enterprises and engineering construction units began to work together to tackle floating wind turbines, floating bodies, mooring systems, long-distance transmission, transportation, installation and operation and maintenance technologies, and have achieved a series of results.

On December 7, 2021, the Three Gorges New Energy invested and built China's first floating offshore wind power platform, carrying the world's first typhoon-resistant floating offshore wind turbine to form the "Three Gorges Leading," which successfully connected to the grid in Yangjiang offshore wind Farm in Guangdong Province.

It marks that China is the first country in the world to have independent research and development, manufacturing, installation and operation capabilities of large-capacity typhoon-resistant floating offshore wind turbines, which has positive significance for promoting the upgrading of China's high-end offshore wind power equipment manufacturing and tapping far-reaching sea wind energy resources.

Figure 3.30 On December 7, 2021, the "Three Gorges Leading," China's first floating offshore wind power platform, was connected to the grid in Yangjiang offshore wind Farm in Guangdong Province.

Subsequently, floating wind turbines such as "Fuyao," "CNOOC Guanlan," "CEIC Gongxiang" and "Mingyang Ocean X" have been commissioned. Among them, "Mingyang Ocean X" adopted a dual-head design, with a capacity of 16.6 MW, which was the largest capacity of floating wind turbines in the world at that time. As of March 2025, in addition to the above five floating wind turbines that have been commissioned, China is also promoting the construction of several floating wind power projects, two of which use units with unit capacity of 16 MW.

Figure 3.31 On December 11, 2024, "Mingyang Ocean X" was been commissioned in Yangjiang of Guangdong province.

While breakthroughs have been made in low-wind speed, high-shear and far-reaching sea wind power technology, China has also made wind turbines operate in high-altitude areas through technological innovation, such as the 100,000 kW guaranteed grid-connected wind power project in Basu County by Datang, which was commissioned on October 31, 2024, with the highest altitude of the wind turbine reaching 5,200 meters. It provides valuable experience and demonstration samples for the development of global ultra-high-altitude wind power projects.

3.2.6 Indispensable Step

In September 2024, Goldwind Technology completed the static load test of its GW147 wind turbine blades at the National Test Center. The blade is 147 meters in length, if installed on the wind turbine, the rotor diameter can reach 300 meters, which is the longest wind blade in the world at that time, and also represents China's wind power entering the era of 300-meter rotor diameter.

In recent years, China's exploration in the field of wind power has reached the world's advanced level. The independently designed and manufactured offshore 20 MW wind turbines have touched the boundaries of existing technologies, and it is urgent to make new breakthroughs in the basic theories and models of wind power, design standards and specifications.

The most basic structure of the exploration material depends on electron-positron collider; the technological progress of aerospace is dependent on large wind tunnel laboratories; the innovation of wind power technology also needs the support of major scientific research infrastructure.

Du Xiangwan, former vice president and academician of the Chinese Academy of Engineering and deputy director of the National Energy Advisory Expert Committee, proposed to crack the design and manufacturing bottleneck of high-power components such as main bearings, blades and gearboxes through public technology experiment platforms, accelerate the test iteration and industrialization process, promote the upgrading of the offshore wind power equipment industry chain, and meet the market demand for high-quality and low-cost supply chains, let China's offshore wind power in the forefront of the world.

Taking wind power blades as an example, from the development history of European and Chinese wind blades, in the 24 years from 1991 to 2015, Chinese wind turbine's power and blades are followers. But by 2017, China had a 171-meter-diameter wind blade, more than Europe's 164-meter-diameter wind blade. In 2019, Europe and the United States launched a larger wind blade with a diameter of 220 meters. In terms of wind wheel diameter, China and Europe almost stand on the same platform.

In fact, in recent years, Chinese machine manufacturers and blade manufacturers have made beneficial attempts in the application of basic theory of blade design innovation and the application of new materials. The industry is also deeply aware that the design, process, testing and verification of large blades and supply chain control all need a closed-loop verification process to provide safe and reliable blades for the market, and this closed-loop verification can feed research and breakthroughs at the basic science level of large blades and help the sustainable development of China's offshore wind power. Based on the actual demand and future expectations, China urgently needs to build a national-level wind power blade public technology test platform.

With the support of the Certification and Accreditation Administration of the People's Republic of China and the local government, and with the encouragement of many experts, China General Certification Center has moved forward step by step, and the guaranteed blade testing center was built in 2010. In June 2018, Certification and Accreditation Administration approved China General Certification Center to build and operate the National Quality Inspection Center in Yangjiang, Guangdong. In January 2021, the National Quality Inspection and Test Center for Offshore Wind Power Equipment in Yangjiang (National Test Center), which is built and operated by China General Certification Center, was put into operation.

Figure 3.32 January 2021, the National Quality Inspection and Test Center for Offshore Wind Power Equipment in Yangjiang, operated by China General Certification, was commissioned.

The National Test Center is the only national-level public technology research and development experimental platform for offshore wind power equipment in China, and the world's largest blade testing laboratory when it is put into operation.

In May 2022, the first 102-meter blade independently produced in China completed testing projects such as weight center of gravity, natural frequency, static test, fatigue test, and static force after fatigue at the National Quality Inspection Center. The blade was designed and manufactured by Sunrui to match a 10 MW offshore wind turbine.

In September 2024, Goldwind Technology completed the static load test of its GW147 wind turbine blades at the National Test Center. The blade is 147 meters in length, if installed on the wind turbine, the rotor diameter can reach 300 meters, which is the longest wind blade in the world at that time, and also represents China's wind power into the era of 300-meter rotor diameter.

Figure 3.33 Goldwind Technology completed the static load test of its GW147 wind blade at Yangjiang National Inspection and Testing Center.

> *The six-degree-of-freedom experimental platform for the electrical and dynamics of 40 MW wind turbines will become a major scientific research infrastructure supporting the R&D and innovation of China's next-generation wind power technology.*

He Dexin also hopes that the government, institutions and enterprises will strengthen joint construction and cooperation, accelerate the construction of a full-size ground test platform for the transmission chain and a high-power component test laboratory, meet the needs of full-size multi-degree of freedom tests for units above 15 MW, and solve the technical problems of key components. Therefore, in addition to strengthening the testing capacity of ultra-long blades, China General Certification Center is also stepping up the construction of large-scale wind turbine transmission chain full-size ground test platform.

Figure 3.34 Qin Haiyan delivering a speech at the 2024 Shantou International Wind Power Technology Innovation Conference.

On December 4, 2024, at the 2024 Shantou International Wind Power Technology Innovation Conference, the establishment of the cooperative innovation cooperation organization and the signing ceremony of the Cooperation Memorandum of "40 MW-class wind turbine Electrical and Dynamics Six-degree-of-freedom Experimental Platform" were held. China Resources New Energy, China General Certification Center, Shanghai Jiao Tong University, South China University of Technology, Shantou University, Jinan University, Golden Wind Technology, Electric Wind Power, Mingyang Group, Envision Energy, CRRC Zhuzhou, Sany Heavy Energy, Yunda Wind Power, United Power, Dongfang Electric, CSIC HZ Windpower, DLJ, CRRC Yongji Motor, CRRC Zhuzhou Motor, Shanghai Electric Group Shanghai Motor Factory, LYC representatives of a total of 21 institutions signed a memorandum of cooperation.

The six-degree-of-freedom experimental platform for the electrical and dynamics of 40 MW wind turbines will become a major scientific research infrastructure supporting the R&D and innovation of China's next-generation wind power technology. It is a huge complex equipment composed of more than 10 advanced subsystems, such as very low speed and ultra-large torque drive motor, super-heavy and high-frequency five-degree-of-freedom loader, high-capacity and high-bandwidth power grid simulator, distributed strong real-time heterogeneous measurement and control network, power hardware in the loop system, and digital twin system.

Figure 3.35 December 4, 2024, Establishment of the collaborative innovation organization for the "40 MW-class Wind Turbine Electrical and Dynamic Six-Degree-of-Freedom Experimental Platform."

In order to cover the test needs of 30 MW wind turbines in the future, the six-degree of freedom experimental platform for 40 MW wind turbines has challenged multiple parameter indicators that break the limits of existing technologies for the first time in the world. It includes 75 MNm level torque, 13 MN level thrust, 120 MNm level bending moment, and 120 MVA level power grid simulator, representing six degrees of freedom loading capacity. Among them, whether it is 75 MNm level torque, or the size of the drag motor that generates the torque, it is close to the world's largest single capacity 1 million kilowatts hydropower unit installed in Baihetan Hydropower Station. 13MN's powerful thrust is derived from a large hydraulic mechanism, which is equivalent to the combined engine thrust of 10 Boeing 747 passenger aircraft at full takeoff. The capacity of the 120 MVA power equipment is equivalent to that of a large substation at the district and county level, which can provide electricity to 10,000 households at the same time. In order to carry this huge equipment, it is necessary to pour a huge reinforced concrete base 60 m long, 20 m wide and 10 m deep under it.

Figure 3.36 40 MW-class Wind Turbine Electrical and Dynamic Six-Degree-of-Freedom Experimental Platform.

Relying on its own accurate and controllable simulation of all working conditions, sensitive stress acceleration and other capabilities, the electrical and dynamic six-degree of freedom experimental platform for 40 MW wind turbines will play a strong leading advantage in the design verification, reliability assessment, innovative technology maturity improvement and other new model research and development, supporting China's wind turbine technology leading the world. The experimental equipment can also be used for the exploration of real physical boundaries, the verification and correction of theoretical models and other basic science problems in the industry, which is expected to break through the theoretical basis of large-scale mechanical electrical and dynamics, and become a major national basic research tool.

Looking back on the development of China's wind power for half a century, it is so passionate and unruffled, it is so difficult and invincible, it is so alone and together. In the history of the development of wind power in the world, by stepping out of a unique and atypical development path, China has created a typical success story worthy of praise! It started from a small unit simply to provide lighting for small islands and pastoral areas yurts, all the way through wind and rain and all the way through the rainbow, with the pride of being placed high hopes, but also the bitterness of being criticized, fortunately, those who love it do not give up or abandon it. Over the years, it has seen wind power become a cost-effective source of energy and is at the forefront of achieving China's strategic goal of "carbon neutrality."

About the Author

QIN Haiyan is Vice President of World Wind Energy Association (WWEA), General Secretary of Chinese Wind Energy Association (CWEA), and Director of China General Certification Center (CGC). He has 30 years of experience in testing and certification and has long been engaged in technology research and evaluation, international promotion and application in wind energy, solar energy and other fields. He has presided over and participated in many national 863 program projects, national science and technology support projects and international cooperation projects. At the same time, He is deeply involved in the formulation and revision of China's Renewable Energy Law and the Renewable energy development Plan of the 13th Five-Year Plan and 14th Five-Year Plan, and is well versed in the development policies and technology application of the renewable energy field.

Chapter 4

On the Path to Solar Power Grids

Jürgen Sachau

Université du Luxembourg, Luxembourg
University of Kassel, Hessen, Germany

juergen.sachau@uni.lu

4.1 Introduction

Given the rapid progress in the feed-in of renewable energies, coupled with thermal energy and electrification of mobility, the accommodation capacity of the power grid structures, with their bottlenecks, is becoming increasingly critical. While existing grid structures only slowly evolve, the requirements emerging are ranging from grid stability further to intraday buffering and to bridging long-term energy surplus and gaps. In order to maintain supply security and open market access in the multilevel structures down to distribution grids, it becomes evident that congestion at utility-grid interconnections requires distributed storage, to shape peaks and ramps, to balance fluctuations and to shift energy throughput to times of low grid usage. The flexibility needed for the coordination of fluctuating power, demand side management and sector coupling is provided by interaction via local buffering and local market demands.

The Modern Energies
Edited by Wolfgang Palz

ISBN 978-981-5352-02-3 (Hardcover), 978-1-003-79730-2 (eBook)
www.jennystanford.com
DOI: 10.1201/9781003797302-4

In multilevel grid structures, temporal and spatial balancing are complementary. In aggregated areawide overall views, they exhibit and also quantify long-term progress in grid integration of renewable power generation and sector coupling. With regard to techno-economic viability and policy effectiveness, competing paths towards solar power grids are combining the rapid increase in using all forms of solar energy with a mix of storage technologies and flexibility in surplus and deficit handling.

Conventional power grids operate based on the approach to meet distributed electricity demand by accordingly planning, operating, and controlling power generation. Beyond that, the solar power grids evolving are characterized by interaction among increasing numbers of various actuators for electricity generation, storage, and consumption. This evolution occurs in an environment where solar power and storage costs are rapidly decreasing, expanding electrical energy usage into new sectors towards electrification of the whole energy system.

4.2 The Era of Digital Micro- and Power Electronics

In the 1980s, the advent of microprocessors revolutionized digital signal processing and control of electric drives. At Prof. Leonhard's control engineering institute at TU Braunschweig, we pioneered advanced control methods using handcrafted boards and bare metal programming. Along with switched power devices as the first digital actuators, this technology was soon applied to speed-variable wind power generation as well. The expertise was then transferred to the University of Kassel, where my team prototyped a first processor board that, once operational, was forwarded to a colleague of Kassel's first PhD generation who implemented a MS-DOS operating system. For the former lab engineers of Prof. Kleinkauf, founding SMA, this marked the beginning of their industry PC development with initial niche automation applications, gaining experience in manufacturing of digital boards.

In between, with my university team, we engineered the overall control design with a digital twin. Industrial boards had become available, still expensive like the first personal computers. For hard-real-time control, we adhered to fast and reliable bare metal programming and hardware-scheduling without informatics overhead. The lab prototype of the first battery inverter-generator forming a stand-alone grid, was implementing the innovative digital voltage-vector control, which led to my PhD. Its rapid disturbance compensation allowed a reduction in the rotating masses, for the same frequency deviation tolerable, down to a fifth.

A real solar enthusiast, SMA's main founder Günther Cramer continued to pursue his vision and firm commitment. He managed to keep SMA in the game over a decade with early midsized power plants. Our cooperation included the first battery-buffered stand-alone grid, demonstrated with wind generators in the 1987 Cape-Clear Island EC project. At the same time, the PV business slowly started, supported by initiatives such as the 1000-Roof Program. However, the first industrial PV inverters faced significant challenges and frequent failures, marking the start of a steep learning curve in this nascent field.

4.3 From Modular Systems and Reliable Inverter Technology to Mass-Manufacturing

In the final decade of the millennium, digital power rapidly evolved, driven by microprocessor-based digital control and switched power electronics. This enabled a pivotal innovation, promoted by Wolfgang Palz's EC-JOULE program and coordinated with EUREC-Agency. Introduced via my systems engineering department at Prof. Kleinkauf's spin-off Institute for Solar Energy Technology, the project proposal focused on renewing the state-of-the-art plant technology. The goal was to create a comprehensive technology, based on grid-compatible AC-coupling of all components. Joining forces with WIP at Munich, ITER at Tenerife and CRES, Athens, it was to address PV and wind energy usage across diverse applications from single-loads—such as desalination and water pumps—to island grids and weak grid-coupled utility-scale plants.

As a key enabling innovation, this required advanced auto-sync inverter control for versatile AC-parallel operation to be developed. Our previous digital control of voltage-vector amplitude and phase could be extended to cooperative voltage and frequency control, akin to the way conventional large generators synchronize. Adding innovative phase damping, intellectual property was assigned to Alfred Engler, one of the project's PhD collaborators. Modular systems technology marked a shift from engineering individual plants to designing modular systems that could be easily configured, combined and extended with grid-compatible components. Now, multimegawatt families of power plants, with AC-coupling of wind and PV generation, storage and backup, should become both reliable and economically feasible.

Figure 4.1 1995 Sunny Boy Inverter with digital signal processing and wireless communication boards stacked.

In parallel, the national project Modular PV Systems Technology explored the potential of PV installations for the whole range of power, needed in the built environment and further to utility-scale plants. Weighing both mechanisms of cost reduction—economy of size and economy of scale—was pointing at an overall cost minimum. It combined a small range of larger single inverters and a high number of small-power AC-parallel inverter groups, covering the whole spectrum of building-integrated PV.

In the early 1990s, with its high module costs, the READ initiative of Wolfgang Palz joined renowned architects under the lead of Norman Foster, Richard Rogers, Renzo Piano, Thomas Herzog and Stefan Behling. Their 1996 charter focused on that half of EU energy consumption attributed to buildings. Signed by 27 prominent architects, it emphasized the importance and potential for integrating PV into architectural design. This was much eased by covering PV-active parts of the building hulls in a modular way, with different orientations and surfaces of PV. Within a decade, AC-modular systems became the standard technology. This led to the development of string inverters, extending PV cell-serial connection to strings of modules, with the inverters' AC-parallel operation already proven, likewise working for battery inverters. In a joint research-industry project with Kassel-based SMA, our team systematically investigated the variety of inverter topologies in view of efficiency, cost and reliability and then prototyped digital control and power electronics. To prevent damage to the power electronic switches, current control quality and voltage spike reduction turned out to be the main challenges. This resulted from SMA's first experiences with their solar inverter for the 1000-roofs program, dating back to 1991. Their industrial focus now was on adding communication and informatics complexity and ramping up series production in reasonable quality, where they could rely on previous manufacturing competence in on-board power supply of train wagons for harsh environmental conditions.

In the new millennium, the German 100,000-roofs program for interest-free loans, pre-financed by the state bank KfW, could rapidly be implemented with versatile string-inverter configurations by qualified electricians, overcoming previous bottlenecks of engineering capacity. It had already been well prepared for years by Hermann Scheer and his like-minded politicians from several parties, also aiming at cost-covering feed-in tariffs. Both instruments reinforced each other, when the revised German Renewable Energy Sources Act (EEG) from 2000 was extended to obligatory grid connection and priority access. With 50 cents/kWh for PV not far from cost-coverage, it first introduced feed-in tariffs for renewable energies nationwide.

Figure 4.2 Berlin Railway Station, 1700 sqm PV roof with nearly 100 string inverters, 2003.

Taken over by more and more countries as an outstanding instrument for market entrance of renewable energy, it kick-start the market entry of PV into the solar age. The first decade saw a hundred-fold increase in annual installations, enabled by rapidly upramping string-inverter production and installation. This development converted SMA into a solar company, world leading in PV inverter manufacturing, with their workforce expanding to over 5,000 employees. String inverters revolutionized PV system integration by adopting the inherent modularity of photovoltaic power generation itself. They enabled large-scale deployment across diverse environments without requiring elaborate engineering solutions. This paved the way for large-scale adoption of PV technology within built environments, an industry standard for scalability, fostering sustainable architectural practices across Europe and beyond. A prime example was the EC demonstration project at Berlin's new central railway station.

Its iconic daylighting PV roof represents a mid-power installation designed for minimal operation and maintenance (O&M). With restricted access above the rails—limited to one hour per night—fault detection relies on centralized monitoring via power-line communication, locating faulty strings isolated from AC-parallel operation through their string inverters. This brings PV-plant performance detriment to a low percentage range and enables well-timed targeted repairs.

Planning and safe and secure installation had gone into the hands of craftsmen, qualified as solateurs within days at facilities like SMA's Solar Academy—a fully energy-autonomous building designed by renowned solar architect Manfred Hegger. His students of TU Darmstadt won the Solar Decathlon 2007 of container-shipped solar houses, built up in Washington, D.C. He spearheaded transformative projects such as converting Hamburg-Wilhelmsburg's concrete bunker, which after WWII withstood demolition, into an "Energy Bunker" featuring a solar roof, south-facing solar façade, visitor center with a café and viewing platform. Like other old military buildings, he reshaped it with a solar hull, he converted this relic of war into a solar peace landmark.

4.4 Know-How Extension and Market Opening for Utility-Scale PV

AC-modular systems technology was further advancing in a number of subsequent European and national RTD projects, also resulting in the foundation of the international DERLab association with their Kassel headquarter. Companies, utilities and research institutes joined forces to broaden the engineering know-how through coordinated RTD works, prototyping and testing, to cope with the challenges arising from bidirectional power flows and strong distributed actuator power in the grids. This covered adapted protection, grid observation, improving stability and reducing harmonics as part of the whole technology know-how required for large-scale integration of digital power. In Luxembourg, this was aligned with its utility CREOS and their grid infrastructure.

Figure 4.3 Tenerife PV Field, ITER 2007.

To realize the vast potential, the distribution system infrastructure is required to

- reinforce grid usage and provide flexibility
- reduce congestion and increase throughput

with the grid operators responsible to maintain reliability, safety and security in a system context with an increasing number and variety of power actuators to be accommodated.

In the meantime, the German boom that had made it a world leader in PV and wind power, found its end when the conservative/liberal government 2012 set a rigid upper limit for new PV power capacity that would receive feed-in tariffs. It was especially PV that endangered mid-day profitability of conventional peak-power plants. Abruptly the new cap truncated support for PV business and its rapid job growth, leading to a number of company bankrupts and over hundred-thousand jobs lost. From the global perspective however, the market entrance gap from high prices at low volumes towards low prices at high volumes was closing. With the PV costs down to about 1/3 during the first decade, larger power plants and central inverters had now become attractive for southern European and global markets.

An early commercial plant in the MW range, built and operated by Tenerife's research institute ITER, brought together investment from PV enthusiasts and Tenerife citizens. Backed by their on-site engineers and technicians, its long-term reliable operation relied on AC-modular multi-inverter structure.

In the second decade, the international utility PV market, led by China was taking over the main growth in installed PV and inverter power. As string inverters in their economy of numbers were not yet coping with the decrease of PV-module costs, SMA like other companies launched container inverters for large-scale utility feed-in. Soon after competing in the leading Chinese and other world markets, several European industry inverter producers discontinued their PV activities, not being sufficiently competitive and also facing problems to cope with weaker grid extensions, especially stable parallel operation, reactive power-, voltage- and harmonics-compensation.

For ever larger PV plants, in the course of longer operation, however, the steady need for qualified operation and maintenance personnel became pressing. To avoid serious performance gaps, it is inevitable to locate module faults in hectare-sized plants, often at remote locations in reasonable time. This led to new constructions of large-power inverters, again towards fine-grained power conditioning. High numbers of low-power inverter slices—like string inverters—now considerably advanced in their economies of scale. They aggregate to packages with auto-parallel operation, with possible intermediate levels and voltage increase up to medium and high voltage. For the densely packed large numbers of inverters, lowering heat dissipation from losses and self-protection of power switches are key for low maintenance needs. Further, even lower manufacturing and operation fault rates became important. Based on experience from mass production of ICT equipment with strict manufacturing and testing standards, manufacturers selling millions inverters globally like Huawei, reached failure rates in the lower per mille range. Quality products have to pass type approval and testing—similar as for modules where performance guarantees over more than 20 years are common—for accelerated humidity and temperature cycling in climate chambers extended to dust, lightning and explosion tests.

Again, communication from large numbers of parallel inverters, fine allocated to PV arrays, to centralized control centers, allows for easy failure detection and encapsulation, easing component replacement in large PV plants. Further, complementary to advanced ICT and power electronics development, companies need to provide engineering capacities to cope with PV grid interface challenges and adapt control software for PV and battery units, same as for variable speed generators of wind and hydropower plants.

4.5 Fully Balanced Solar Power Grids and the System Value

Reinforcing variable solar power generation with buffering is crucial for supporting decentralized grids. This approach allows for energetic balancing of fluctuating power, demand side management, and the integration of transport and thermal energy sectors. These systems interact with each other and with markets, stabilized by distributed buffering. Considerable balancing capacities become necessary, once the energy-share of intermittent solar PV and wind usage is rising beyond ranges of 30–50%. This goes well along with the utilization reserves of distributed grids. Typically, they can be increased by factors of two to three until operating close to power-strain limits of the grid infrastructures.

Figure 4.4 60 MW Gannawarra Solar Farm with Battery Container Field, Australia 2018.

Storage capacities are by far led by distributed batteries and—more centralized—pumped hydro storage plants. Decreasing costs of intraday battery-buffered PV installations arrive at the cost of PV-only installations with a delay of 2–3 years. For pumped hydropower, Australian plans from their national university, Canberra, in their survey and map of suitable storage sites, resulted in a national potential and a global potential as well, two orders of magnitude higher as needed for fully balanced solar-electric grids. End of last decade, their costs of hourly balancing were about half of the PV and wind generation costs.

With the climate changing, it is getting ever more irresponsible to operate conventional thermal base-load power plants, vaporizing large amounts to cool off the major part of the thermal energy as waste heat. They often depend on the exploitation of external raw materials, produced under conditions impossible in Europe. Instead, many hydro power plants can be enhanced with storage basins. Like the glacier effects, delayed water release can serve as efficient countermeasure against droughts and floods. Europe has thus aligned hydropower usage with the EU water-framework. The Alpine countries, foremost Austria and Switzerland, make extensive use of their hydropower potentials—the Swiss energy strategy foresees new projects, most of them raising the heights of the existing reservoir basins, some by more than 20 m, allowing them to fully balance their electricity-needs, to achieve their target of 100% autonomous renewable energy supply. As of 2025, pumped hydro energy storage globally performs roughly 5 percent of electricity storage. The global atlases of the Australian National University exhibit vast potential worldwide for small water and land requirements and low cost.

Huge utility-scale PV plants, often in remote regions like China, are increasingly paired with large pumped hydro facilities to also provide long-term balancing. Complementary storage technologies, among others, based on solar hydrogen and synthesis of energy-rich hydrocarbon fuels have gained significant traction, with Europe's largest shipping company committed to converting all its vessels to climate-neutral operation by 2030.

Addressing the needs for energy balancing requires integrated planning and projecting, taking into account all limitations in the distributed grid infrastructure. The system value of flexibility, with balancing as the main energy component, is to be observed spatially distributed and across different time frames. Performance monitoring, as established in the guidelines of JRC and IEC standards, is extending to AC-coupled storage, with power flow decomposition into directed power flows. Utility interconnections are thus characterized not only by power-duration curves but also by energy-duration curves, detailing energy-shifts over time-ranges. This approach recursively applies for larger grid areas, employing unified performance evaluation and analysis of operation data. Normalized to system size and across different time frames, these metrics provide the basis for monitoring storage performance, provision of flexibility, and contracting of ancillary services.

With SMA in Germany and Fronius in Austria, the inverter manufacturing base faces intense global competition and cost pressure, leading to significant outsourcing of manufacturing to countries like China. As a result, the market share of European inverters has decreased to around one-fifth.

Figure 4.5 Pumped Hydro Power Plant Limmern, Swiss Alps.

Standard monitoring data from grid and storage sites is providing valuable experience to verify and improve techno-economic performance projections, as from Open Energy Model Initiatives. On the basis of shared data ownership, comparative data evaluation is improving good practice, with automated processing rapidly evolving. Beyond support for project design and investment decision, this allows for the pursuit of overlaid economic viability and policy effectiveness.

4.6 Inverters at the Core of Engineering Wide-Area Solar Grid Infrastructures

PV and battery inverters, along with inverters for electrochemical and speed-variable electromechanical conversion, play a crucial role in the grid interface. They are essential for ensuring supply quality, reliability and techno-economic integration. In Europe, after years of rapid growth, the solar sector had experienced an intermediate slowdown. However, solar PV remains the major component of installations, with operating solar power exceeding 300 GW. At the heart of solar power grids, the manufacturing of PV and battery inverters, along with related grid-integration engineering, forms the dominant European solar business segment. By the mid-decade, EU inverter production is approaching 100 GW, still a backbone of solar employment, with international leaders like

Beyond the recent boom in building-integrated solar, power grids worldwide are struggling to cope with solar growth. Utility-scale solar installations are increasingly restricted by grid and flexibility bottlenecks, which are systemic barriers that need to be overcome. To achieve goals related to competitiveness, climate, and energy security, regulatory stability and sustained structural financial support are crucial. This includes improving flexibility and grid development to enhance electrification, e.g., in the EU from about 23% to 35% by the end of the decade, and to about one-half of expected total energy usage by mid-century.

As key components, inverters still bear considerable potential for both cost reduction through economies of scale and improved integration through communication and fast cooperative control. The decrease in the price of inverter technology has well kept pace with photovoltaics. For utility-scale projects, they are accounting for 5–10% of the costs for modules, inverters, and their mechanical and electrical installation, which together make up roughly half of the total.

They offer full self-diagnostics and self-protection, allowing for fine detection of faults in large utility installations. With mass production of low-failure single-slice string inverters and their dense packaging for inverter groups, they remain cascadable in failsafe auto-parallel operation for AC-power aggregation and conversion to higher voltage levels. Over two decades of increasing efficiencies, losses have come down from above 10% to the range of 1%. Faster switching semiconductors lead to smaller passive components. Emerging technologies like GaN-powerchips may also allow for integration of logical and control functions.

Coupling with battery inverters and leveraging experience from topologies applied in other fields, such as multilevel inverters, offers further potential to support entirely solar-electric power grids. This addresses reinforcement of existing infrastructures as well as evolving new grids, their extensions or compounding into larger structures, on the foundations extended for phase-stable multi-inverter control, identification, protection, auto-curtail control and self-organizing actuator cooperation in markets. The capabilities of inverters for power generation, storage, and controlled usage lay the ground for solving grid engineering problems while keeping infrastructure expenditures reasonable. The progressive large-scale exploitation of variable PV and wind power feed-in depends on associated buffering support and long-term balancing capabilities in their grid context. The evolving systems technology of solar power grids requires integral planning and engineering of solar power, balancing and grid cooperation, structurally transforming conventional grid structures and exploring new ones.

About the Author

Jürgen Sachau is a former EC official and full professor of Universities of Kassel and Luxembourg. Studying electrical and control engineering at TU Braunschweig, he joined Prof. W. Leonhard's pioneering digital control in the early 1980s, implementing digital grid damping control of a synchronous generator. Under his supervision, he continued with his University of Kassel team to prototype battery inverter coupled island supply, the novel vector control leading to his PhD. In the

1990s, his systems engineering department at Germany's newly founded Solar Energy Technology Institute, as part of EUREC-Agency, laid the foundation for AC-modular wind and PV systems with lab prototypes, including the first string-inverter, initiating large-scale building integration. Prof. Sachau served the EU at DGs Research and JRC, responsible for some 50 RTD projects and for monitoring the EU PV demonstration, also extending their guidelines for grid-coupled storage. At Luxembourg, his department, with the international DERLab association, laid the engineering foundations towards fully solar power grids. He was editor for the *European and International Transactions on Electrical Power* and editor-in-chief of the *International Journal of Solar Energy* and joined IEA, IEC and EC committees and the Norman Foster solar-prize jury. His works on whole-range dynamics in multilevel grid structures focus on reliable systems & control engineering, foundations and tools for solar power grid integration from planning and projecting to protection, economic operation and control.

Chapter 5

The Evolution of Biomass and the Role of EUBCE

Angela Grassi
ETA-Florence Renewable Energies, Florence, Italy
angela.grassi@etaflorence.it

For more than 40 years, the European Biomass Conference and Exhibition (EUBCE) has provided a unique perspective on the evolving trends in bioenergy. It has served as both a retrospective on the development of the biomass sector and a forward-looking forum for experts in research, industry, and policy. As an organizer, I have had the privilege of witnessing this evolution firsthand, drawing from the extensive legacy of EUBCE's proceedings—thousands of pages of printed books and digital documents, carefully preserved in the archives at ETA-Florence Renewable Energies. Delving into these records is like embarking on a journey through time, offering a glimpse into how the future of biomass was once envisioned.

The Modern Energies
Edited by Wolfgang Palz

ISBN 978-981-5352-02-3 (Hardcover), 978-1-003-79730-2 (eBook)
www.jennystanford.com
DOI: 10.1201/9781003797302-5

Figure 5.1 The Founders Wolfgang Palz and Giuliano Grassi (top) and Angela Grassi (bottom, center).

5.1 The Early Years: Laying the Foundations for Biomass Research and Policy

Since the inception of renewable energy policies and research programs in Europe, EUBCE has been instrumental in fostering knowledge exchange. The first "International Conference on

Biomass," held in 1980 in Brighton, UK, was organized by the Commission of the European Communities, DG Research, Science, and Education, in collaboration with the UK's Natural Environment Research Council. These early conferences had three key objectives:

1. Assisting the European Commission in defining its R&D programs.
2. Facilitating knowledge-sharing among national research initiatives to avoid redundancy.
3. Promoting international cooperation with both industrialized and developing countries.

At the time, renewable energy technologies were in their infancy, and the potential of biomass as a viable energy source was not widely recognized. The primary driver for diversifying energy sources was ensuring Europe's energy security, with nuclear energy still playing a dominant role. As noted by the cochairs of the first conference, Philippe Chartier, Scientific Director of the French Agency for Energy Management, and David Hall, University of London, King's College, United Kingdom:

> *"For some European countries, awareness of the promise of biomass has been slow to develop. Biomass has been regarded as a marginal source of energy that could at best make only a small contribution to national energy supplies. This conference made clear that biomass has much more than a marginal potential... The use of biomass as a source of energy in the EC could in the medium term meet 5% of the Community's energy requirements and would have a positive effect on employment, the pursuit of regional policy measures, and the balance of payments."*

Comparing this statement to today's reality highlights the progress achieved.

In 2022, Bioenergy Europe reported that bioenergy accounted for 55.7% of the EU's renewable energy consumption, representing about 13% of gross final energy consumption. This sector also supports substantial employment; IRENA data shows over 3.5 million global jobs in biofuels, biomass, and biogas as of 2021.

5.2 From Research to Industry: The Evolution of Biofuels

The advancement of biomass as a key energy source was made possible through sustained efforts in research and development, alongside supportive policies at national and European levels. Throughout this evolution, the European Union (and its predecessor, the European Community) played a fundamental role in funding and coordinating these efforts.

Early research in the 1980s focused on biofuels for transportation, particularly methanol production from wood as a fuel additive (MTBE). The 1980 conference proceedings stated:

> *"At present time, methanol as a liquid fuel has better prospects than ethanol in developed and developing countries... The advantages of methanol derive from the following considerations: gasification of lignocellulosic biomass is today more efficient than hydrolysis and fermentation; cost analysis and energy balance are more favorable."*

However, concerns about methanol's toxicity and environmental risks led to a shift in focus towards ethanol and biodiesel. While these fuels were initially considered more suitable for developing countries, they eventually became mainstream in Europe.

By 2000, EUBCE's conference in Seville highlighted growing interest in lignocellulosic ethanol, although progress remained within the research domain, primarily in North America. Four years later, the Rome conference showcased tangible technological advancements and increased industrial commitment to second-generation biofuels. Companies began announcing the first semi-commercial plants.

In 2013, at the Berlin conference, Beta Renewables introduced the Crescentino Biorefinery in Italy—the world's first commercial-scale lignocellulosic ethanol plant. By 2015, UPM Biorefining had launched its Lappeenranta biorefinery in Finland, producing 100,000 tons per year of renewable diesel from forestry residues. That same year, the Flightpath Initiative, a partnership between the aviation industry and the European Commission, reported over 1,500 commercial flights powered by bio-kerosene blends.

At the time, the advanced biofuels industry was in its early stages, but these milestones highlight how breakthroughs from the early 2000s have evolved into proven industrial technologies. Acknowledging these achievements is essential for building the consensus necessary to establish a stable European policy framework for bioenergy.

5.3 The Luminous Vision of 2019

Fast forward to 2019, when EUBCE set the stage for a new era: Experts demonstrated how mature biomass conversion technologies were producing power, heat, and transportation fuels at an industrial scale. Advanced biofuels were no longer a distant vision but a tangible reality, helping to decarbonize long-haul flights, shipping, and heavy road transport.

5.4 The Digital Leap in 2020

The 2020 edition, held online due to the global pandemic, marked a pivotal shift towards digitalization. Discussions focused on optimizing biomass conversion technologies while addressing challenges such as land-use conflicts and environmental trade-offs. The emphasis was on advanced biofuels derived from waste, reinforcing the rapid transition from research to industrial application.

5.5 Technology Meets Policy: 2022–2024

In 2022, EUBCE highlighted the intersection of technology and policy, showcasing how digital tools—such as artificial intelligence—were refining biomass conversion processes. The following editions in 2023 and 2024 continued this trajectory, presenting commercial-scale biorefineries producing lignocellulosic ethanol and innovative pilot plants converting forestry residues into renewable diesel. These developments underscore how decades of R&D, coupled with strategic policy support, have transformed biomass into a cornerstone of Europe's circular bioeconomy.

5.6 Advancing Biofuels and Biochemicals: 2025

The 2025 conference showcased the pivotal role of biomass in Europe's climate strategy, highlighting its versatility for energy, fuels, materials, and carbon removal. Biofuels were positioned as the central pillar of transport decarbonization, particularly for aviation and maritime, while also contributing to energy security, regional development, and land restoration. Scaling up will benefit from de-risking measures, innovative agricultural practices, and new feedstocks from waste and marginal land. Advances in biorefineries, CO_2 valorization, and bio-based products confirmed that biomass is not only ready to expand but also essential in driving the transition to a net-zero economy.

Sustainable Aviation Fuels were recognized as essential to aviation's global decarbonization. Industry stakeholders called for coherent long-term policies, harmonized standards, and investment support to accelerate deployment. Diversification beyond HEFA and the development of e-SAF will unlock the sector's full potential and ensure aviation contributes decisively to the net-zero transition.

5.7 Advancing Sustainable Aviation Fuels: 2025

At the EUBCE 2025 edition in Valencia, a High-Level Round Table on Sustainable Aviation Fuels (SAF) emphasized aviation's unique opportunity to act as a global leader in decarbonization. Industry stakeholders highlighted the growing momentum of SAF production and the importance of coherent long-term frameworks, global harmonization of standards, and diversification beyond HEFA pathways towards Alcohol-to-Jet, new energy crops, and e-SAF. With the right policies and investments, the sector is well positioned to scale rapidly, secure future supply, and ensure aviation makes a decisive contribution to global climate goals.

Figure 5.2 EUBCE 2025 special event on SAF.

5.8 The Future of EUBCE: Bridging Research and Industrial Implementation

Figure 5.3 Announcing EUBCE 2026.

The EUBCE will continue to focus on closing the gap between research breakthroughs and industrial-scale implementation. With the ongoing energy transition and the push for decarbonization, biomass will remain an essential component of Europe's energy mix, contributing to climate goals and energy security.

5.9 About EUBCE

EUBCE is the most significant international conference dedicated to the biomass sector, complemented by a technology exhibition. For over 30 years, it has served as an annual gathering for biomass experts across research, industry, and policy. The conference covers cutting-edge technological advancements, policy frameworks, and long-term strategic potential, acting as a vital interface between science, industry, and policymakers.

The program is coordinated by the European Commission's Joint Research Centre, with scientific and industrial contributions from a committee of more than 140 international biomass experts, and with Industry Program Support provided by the Circular Bio-based Europe Joint Undertaking.

For more information, visit: www.eubce.com

Figure 5.4 EUBCE 2024.

Figure 5.5 EUBCE 2025.

Figure 5.6 EUBCE 2025 Indian delegation.

Figure 5.7 EUBCE 2025 Republic of Korea delegation.

Figure 5.8 Linneborn Prize 2019.

Figure 5.9 Giuliano Grassi Prize 2025.

About the Author

Angela Grassi is the Managing Director of ETA-Florence Renewable Energies. With a background in economics and political science, she has over 30 years of experience in the renewable energy sector, specializing in international project management, strategic dissemination, and high-level event organization. She has coordinated numerous EU and international projects, leading multidisciplinary teams and overseeing all phases from project definition to reporting, budgeting, evaluation, and final delivery to the European Commission. Angela is also an accomplished author and editor, with publications in national and international conference proceedings, contributions to specialized books, and extensive experience as managing editor of brochures, publications, and scientific proceedings. A recognized leader in knowledge exchange, she has organized and managed major international conferences, exhibitions, workshops, and tailored events worldwide. Since 2000, she has served as lead organizer of the **European Biomass Conference and Exhibition (EUBCE)**—one of the world's foremost events in the sector (www.eubce.com)—and, since

2024, she has coordinated the **Integrated PV Conference (IPV)** (www.ipv-conference.com). She has also served on the Executive Committees of several international conferences, consolidating her role as a key figure in advancing global dialogue, stakeholder engagement, and knowledge exchange in renewable energies.

Chapter 6

What Role for Nuclear Power in the Ongoing Energy Revolution?

Mycle Schneider

Independent international analyst and consultant on energy and nuclear policy

mycle@orange.fr

In 2017, then 91 years old, the great, late S. David Freeman declared "the debate is over" as "nuclear power has been eclipsed by the sun and the wind. These renewable, free-fuel sources are no longer a dream or a projection—they are a reality replacing nuclear as the preferred choice for new power plants worldwide."

Few people knew the energy sector better than the engineer and lawyer Dave Freeman. Appointed Chairman of the Tennessee Valley Authority (TVA) by U.S. President Jimmy Carter in 1977, he subsequently served for two decades as general manager of several large public power agencies, including the Los Angeles Department of Water and Power, the New York Power Authority, and the Sacramento Municipal Utility District.

I had invited Dave Freeman to provide a foreword to the *World Nuclear Industry Status Report 2017* and was absolutely thrilled that he accepted. When I had read his wonderful draft, I still asked

The Modern Energies
Edited by Wolfgang Palz

ISBN 978-981-5352-02-3 (Hardcover), 978-1-003-79730-2 (eBook)
www.jennystanford.com
DOI: 10.1201/9781003797302-6

him whether his "debate is over" statement was not over the top. He maintained every word of it. And, of course, he was right.

When Dave Freeman passed away in 2020, the *New York Times* wrote in a remarkable obituary: "His belief that renewables could become economically competitive with other forms of power generation went from a minority view to a widely accepted one. The United States is on track this year to produce more energy from renewables than from coal."

Since Dave Freeman closed the debate in 2017 until 2023, power generation in the United States from coal dropped by 46 percent, from nuclear by 3 percent, while increasing by 90 percent from non-hydro renewables, mainly wind and solar. In 2024, 93 percent of the capacity added to the U.S. grid came from solar (52%), wind (12%), and battery storage (29%). For 2025, the U.S. Energy Information Administration predicts a 26-percent growth of solar capacity to exceed 150 GW in total.

The most staggering development is certainly in battery storage, virtually inexistent in 2017 and on track to exceed a total of 45 GW in 2025, with over 18 GW to be added during the year. Most of the capacity additions in the United States are now hybrid plants, solar *plus* storage or wind *plus* storage. The complementary design on facility level allows for improved efficiency over grid-based balancing.

And where is nuclear? Since 2017, seven nuclear reactors have been closed in the United States, two started up, and none is under construction. In fact, not a single nuclear construction is currently active on the entire American continent, from Canada to South America.

The global picture is hardly any different. Over the past 20 years, 2005–2024, we have seen 104 startups and 101 closures. However, 51 of the new grid connections were in China, with no closures. In other words, the world outside China saw only 53 startups but 101 closures, a net decline of 48 units. With 411 units operating at the beginning of 2025, the decline has been significant.

Has there been any change in the trend over the past five years that did see steady amplifying trumpeting of a "nuclear renaissance"? Globally, 40 construction starts took place in that period, of which 26 were in China, 1 in Pakistan (by Chinese companies), and the remaining 13 implemented by Russian company Rosatom in Egypt, India, Turkey, and at home. Basically, nuclear construction can be summed up by saying China builds at home and Russia abroad. A lot of talk everywhere, but nothing else on the ground, anywhere.

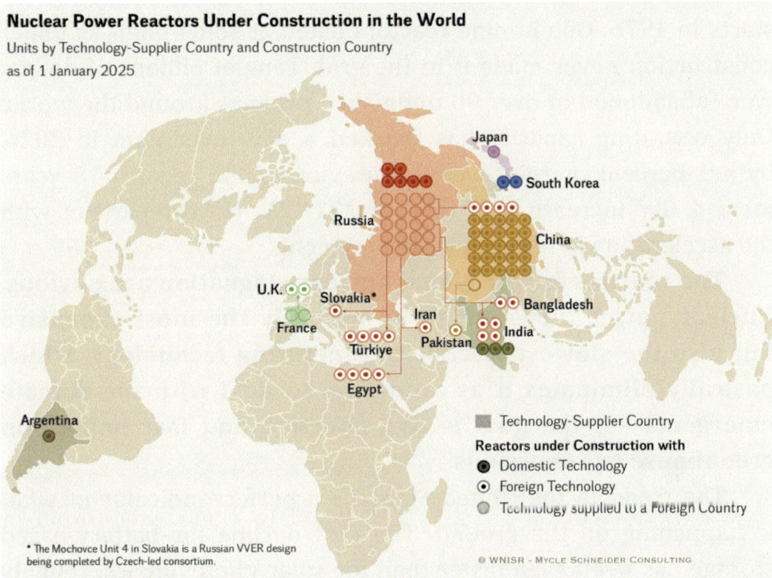

Figure 6.1 Nuclear power reactors under construction in the world (as of January 1, 2025).

While China has been the only country massively expanding its nuclear program over the past two decades, it is dwarfed by its renewable energy expansion. Solar power generation passed nuclear output in 2022 for the first time; in 2024, solar produced already almost twice as many kilowatt-hours as nuclear. No wonder, in the past year alone, China added an estimated 277 GW of solar—more than twice the total the United States installed over the past 40 years—compared to 3.5 GW of nuclear.

So even if in China, officially, nuclear generates seven times more power per installed GW than solar, the nuclear expansion is comparatively slow and it does not contribute more than 4 percent to the total production, with a declining trend.

On a global scale, most of the nuclear power indicators have peaked many years ago.

The historic maximum of operating reactors was reached in 2002, the nuclear share of commercial electricity generation in 1996, new grid connections peaked in 1984–85, the number of units listed as under construction in 1979, and the construction starts in 1976. One in nine reactors listed at some point as under construction never made it to the grid. Tens of billions of dollars were abandoned in over 90 unfinished projects around the world. Only operating capacity has reached a new maximum in 2024, by a 1-percent margin over the previous 2006 record: 18 years for a 4 GW increase to a total of 371 GW. No comparison with the acceleration in renewables deployment.

The reasons for the global nuclear stagnation are obvious. Nuclear power turns out to be not only the most expensive but also the slowest electricity-generating technology, which basically eliminates it as an option to deal with the climate emergency—if the goal is cost-effective and fast impact on greenhouse gas emissions.

The trend in the United States is a perfect indicator of what is happening on the ground. The cost decline for battery-based storage has been even faster than for solar photovoltaics, roughly by a factor of 10 over a decade. The OECD's International Energy Agency has calculated that the Levelized Cost of Energy—the lifetime costs of power generated, including all investment and operating costs—even if "value-adjusted" for estimated additional grid-integration costs for solar *plus* battery storage are already more competitive than coal in China and India, and will be more competitive than gas in the United States by 2030, meaning virtually cheaper than anything anywhere by then.

By 2030, no new nuclear plant will be producing power that is not now under construction. There are only a handful of nuclear builders out there that have completed projects over

the past two decades. They are all state-owned and they have all substantial problems.

- The two major Chinese companies CNNC and CGN are blacklisted by the U.S. Government, which shuts off many companies and countries from any cooperation. Also, China has never built anywhere outside the homeland, if not in Pakistan... where it is at home when it comes to nuclear technology.
- Russia, the only country that is building abroad on a significant scale, is severely hindered by ever-increasing sanctions.
- South Korea's KEPCO has completed one project with four units at the Barakah site in the United Arab Emirates, under unique circumstances. Labor conditions, regulatory oversight, and financial arrangements cannot be compared to any other place. While any information about costs has been classified, the breathtaking debt level of US$144 billion (KRW205 trillion) is an indication that it is not going well.
- France is hosting the largest nuclear operator in the world but which is also struggling with a high debt level of US$62.5 billion (>€54 billion), an aging operating fleet, two units under construction in the U.K. with skyrocketing cost estimates and repeated delays, will not have any new reactor operating in France until the late 2030s, if things go well.
- And then there used to be Westinghouse, which went bankrupt in 2017 over reactor building and has decided not to take that risk anymore. The company will sell design and engineering; the investment risk has to be borne by other companies.

There are no other experienced companies out there. The idea that some startups will solve all the problems of the old established companies is absurd. Building a nuclear power plant is complex, from design to implementation, with operating having its own challenges. This is particularly true for so-called Small Modular Reactors, sold as Small Miraculous Reactors, where promoters have created the public

perception they can be ordered online at Amazon. In reality, none has been built outside China and Russia. None has been certified in the Western world. While in April 2025, the Canadian regulator handed out a conditional construction license to Ontario Power Generation, construction has not started, and the proposed design, the GE-Hitachi BWRX-300, has not even been licensed in the U.S. yet. In fact, the GE-Hitachi consortium has never built any reactor up to grid connection anywhere.

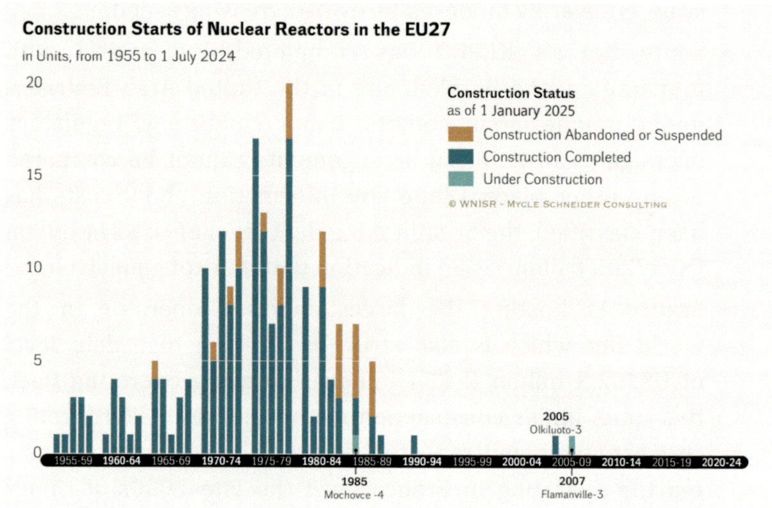

Figure 6.2 Construction starts of nuclear reactors in the EU27 from 1955 to 2024.

The modular design is nothing new. In 2006, and up to construction start in 2013, Westinghouse claimed "by using modular construction methods, Westinghouse and its project partners will be able to build the AP1000 in 36 months." Construction of two units at the V.C. Summer site in South Carolina was abandoned after spending about US$10 billion and nine rate increases for consumers. The bankrupt Westinghouse was taken over by a Canadian holding that has never built a reactor. Two other units were connected to the grid at the Vogtle site in Georgia in 2023 and 2024 after over 10 years of building time, seven years more than planned. The latest reactor startup in the European

Union was in France, where EDF connected Flamanville-3 to the grid 17 years after construction start, 12 years later than planned. Since then, it has been mainly down for unplanned technical reasons.

The nuclear industry has seen a negative learning curve in many countries over the years. Projects are taking longer to build and getting more expensive over time rather than declining in cost as in other industrial sectors. A key reason is that with only five units completed every year over the past 20 years and only 2.5 outside China, it is difficult to learn from experience, from modular production, and competition.

Renewables have a dramatic advantage. Tens of thousands of wind turbines and hundreds of millions of photovoltaic modules are being designed, fabricated, assembled, shipped, installed every year. Manufacturers make huge progress in short periods of time because they can constantly adapt their processes and logistics on the basis of vast experience. This advantage will only grow and worsen the pain of the nuclear industry. It is within the nature of the beast and has nothing to do with ideological pros and cons. It is even outside any considerations of safety, security, and nuclear proliferation, the classical issues of the nuclear controversy.

Unfortunately, few people have understood that "the debate is over." Instead, it has turned into an increasingly fact-free exercise driven by geopolitics, dubious financial interests, voracious appetite for government handouts, propagandists, and some religious believers in 20th-century technology. All of this has little or nothing to do with energy or climate policy, if not this: Politicians have catastrophically failed climate protection policy over the past half century. Rapidly declining weather and climatic conditions are the impeccable demonstration of policy failure. Nuclear power appears to provide a silver-bullet response, if not immediately, then sometime in the future. Until then, some say "drill baby drill" or "dig up that beautiful clean coal."

It is unlikely that the current trend is reversible, even in the world's largest economy. Conservative Texas has built up in a few years 40 GW each of solar and wind, combined with 20 GW of battery storage. Of course, renewables are an industrial reality today on a vast scale. We also know that it is not enough, that sufficiency options and efficiency potentials remain broadly underexploited, and that hundreds of billions in subsidies continue to support fossil fuels. Political and economic leaders need to come back to fact-based decision-making in energy policy. Focus, vision, and investments need to go *now* where they allow for short-term cost-effective results. Promises are powerful, but they are not enough to implement progress. And... Dave was right, of course.

About the Author

(*Photo by Nina Schneider*)

Mycle Schneider works as an independent international analyst and consultant on energy and nuclear policy. He is the initiator, coordinator, and publisher of the annual *World Nuclear Industry Status Reports* and a Founding Board Member and Spokesperson for the International Energy Advisory Council (IEAC). Since 2007, he is a member of the International Panel on Fissile Materials (IPFM), based at Princeton University, USA.

From 2000 to 2010, he was an occasional advisor to the German Environment Ministry. From 1998 to 2003, he was an advisor to the French Environment Minister's Office and to the Belgian Minister for Energy and Sustainable Development. Schneider has given evidence or held briefings at national Parliaments in 16 countries and at the European Parliament. He has advised Members of the European Parliament from four different groups over the past 30+ years. He has given lectures or had teaching appointments at over 20 universities and engineering schools in a dozen countries.

Index